COMPREHENSIVE BIOCHEMISTRY

ELSEVIER SCIENTIFIC PUBLISHING COMPANY

335 Jan van Galenstraat, P.O. Box 211, Amsterdam, The Netherlands

AMERICAN ELSEVIER PUBLISHING COMPANY, INC.

52 Vanderbilt Avenue, New York, N.Y. 10017

Library of Congress Card Number 62–10359
ISBN 0-444-41282-4

With 2 plates, 62 illustrations and 19 tables

PRINTED IN THE NETHERLANDS

COMPREHENSIVE BIOCHEMISTRY

COMPREHENSIVE
BIOCHEMISTRY

SECTION I (VOLUMES 1–4)
PHYSICO-CHEMICAL AND ORGANIC ASPECTS
OF BIOCHEMISTRY

SECTION II (VOLUMES 5–11)
CHEMISTRY OF BIOLOGICAL COMPOUNDS

SECTION III (VOLUMES 12–16)
BIOCHEMICAL REACTION MECHANISMS

SECTION IV (VOLUMES 17–21)
METABOLISM

SECTION V (VOLUMES 22–29)
CHEMICAL BIOLOGY

SECTION VI (VOLUMES 30–33)
A HISTORY OF BIOCHEMISTRY

COMPREHENSIVE BIOCHEMISTRY

EDITED BY

MARCEL FLORKIN

Professor of Biochemistry, University of Liège (Belgium)

AND

ELMER H. STOTZ

*Professor of Biochemistry, University of Rochester, School of Medicine
and Dentistry, Rochester, N.Y. (U.S.A.)*

VOLUME 29   PART B

COMPARATIVE BIOCHEMISTRY, MOLECULAR EVOLUTION

(*continued*)

ELSEVIER SCIENTIFIC PUBLISHING COMPANY

AMSTERDAM · LONDON · NEW YORK

1975

CONTRIBUTORS TO THIS VOLUME

JOZEF DE LEY, D. Sc., Aggr. H.E.

Professor of Microbiology, Laboratorium voor Microbiologie en Microbiële Genetica, Rijksuniversiteit Gent, 35 K.L. Ledeganckstraat, B 9000 Gent (Belgium)

MARCEL FLORKIN, M. D., Hon. D. Sc., Hon. M.R.I., Hon. F.R.S.E.

Emeritus Professor of Biochemistry, University of Liège, 17 Place Delcour, Liège (Belgium)

KAREL KERSTERS, D.Sc.

Associate Professor, Laboratorium voor Microbiologie en Microbiële Genetica, Rijksuniversiteit Gent, 35 K.L. Ledeganckstraat, B 9000 Gent (Belgium)

GENERAL PREFACE

The Editors are keenly aware that the literature of Biochemistry is already very large, in fact so widespread that it is increasingly difficult to assemble the most pertinent material in a given area. Beyond the ordinary textbook the subject matter of the rapidly expanding knowledge of biochemistry is spread among innumerable journals, monographs, and series of reviews. The Editors believe that there is a real place for an advanced treatise in biochemistry which assembles the principal areas of the subject in a single set of books.

It would be ideal if an individual or small group of biochemists could produce such an advanced treatise, and within the time to keep reasonably abreast of rapid advances, but this is at least difficult if not impossible. Instead, the Editors with the advice of the Advisory Board, have assembled what they consider the best possible sequence of chapters written by competent authors; they must take the responsibility for inevitable gaps of subject matter and duplication which may result from this procedure.

Most evident to the modern biochemist, apart from the body of knowledge of the chemistry and metabolism of biological substances, is the extent to which he must draw from recent concepts of physical and organic chemistry, and in turn project into the vast field of biology. Thus in the organization of Comprehensive Biochemistry, the middle three sections, Chemistry of Biological Compounds, Biochemical Reaction Mechanisms, and Metabolism may be considered classical biochemistry, while the first and last sections provide selected material on the origins and projections of the subject.

It is hoped that sub-division of the sections into bound volumes will not only be convenient, but will find favour among students concerned with specialized areas, and will permit easier future revisions of the individual volumes. Toward the latter end particularly, the Editors will welcome all comments in their effort to produce a useful and efficient source of biochemical knowledge.

M. FLORKIN

Liège/Rochester

E. H. STOTZ

PREFACE TO SECTION V

(VOLUMES 22-29)

After Section IV (*Metabolism*), Section V is devoted to a number of topics which, in an earlier stage of development, were primarily descriptive and included in the field of Biology, but which have been rapidly brought to study at the molecular level. "*Comprehensive Biochemistry*", with its chemical approach to the understanding of the phenomena of life, started with a first section devoted to certain aspects of organic and physical chemistry, aspects considered pertinent to the interpretation of biochemical techniques and to the chemistry of biological compounds and mechanisms. Section II has dealt with the organic and physical chemistry of the major organic constituents of living material, including a treatment of the important biological high polymers, and including sections on their shape and physical properties. Section III is devoted primarily to selected examples from modern enzymology in which advances in reaction mechanisms have been accomplished. After the treatment of Metabolism in the volumes of Section IV, "*Comprehensive Biochemistry*", in Section V, projects into the vast fields of Biology and deals with a number of aspects which have been attacked by biochemists and biophysicists in their endeavour to bring the whole field of life to a molecular level. Besides the chapters often grouped under the heading of molecular biology, Section V also deals with modern aspects of bioenergetics, immunochemistry, photobiology and finally reaches a consideration of the molecular phenomena that underlie the evolution of organisms.

M. FLORKIN

Liège/Rochester   E. H. STOTZ

CONTENTS

VOLUME 29 B

COMPARATIVE BIOCHEMISTRY, MOLECULAR EVOLUTION
(continued)

Chapter III. Biochemical Evolution in Bacteria
by J. DE LEY AND K. KERSTERS

Chapter IV. Biochemical Evolution in Animals
by MARCEL FLORKIN

CONTENTS

XI

Chapter V. Ideas and Experiments in the Field of Prebiological
Chemical Evolution
by MARCEL FLORKIN

COMPREHENSIVE BIOCHEMISTRY, VOLUME 29 A
ERRATA AND CORRIGENDA

page ix, item I.2 *instead* of Phylogen, *read* Phylogeny

page 14, line 13 *instead* of Saussure, *read* De Saussure

page 34, line 6, *instead* of amin acid, *read* amino acid

line 22, *instead* of tein synthesis, *read* protein synthesis

page 57, line 4 of section 23,

instead of chlorophylophores, *read* chlorophyllophores

line 5 of section 23,

instead of chlorophylophages, *read* chlorophyllophages

page 60, line 26, *instead* of polychetes, *read* annelids

page 64, line 22, *read*:

Fitch[126a] and Fitch and Margoliash, who have devoted to the implications of the homology concept much penetrating thought, distinguish *orthologous* homology from *paralogous* homology (the original coinage of the terms *paralogous* and *orthologous* was in the paper of Fitch[126a]):

page 78, line 28, *instead* of biosyntagm, *read* biosyntagms

page 86, line 10, *instead* of syntagm, *read* biosyntagm

page 95, line 11 from below, *instead* of the gene-producing clupein Z, *read* the gene producing clupein Z

line 9 from below, *instead* of by a crossing-over with a non-allele gene, *read* by a crossing-over (see Fitch[164a]).

page 120, insert reference 126a:

126a W. Fitch, *Syst. Zool.*, 19 (1970) 99.

page 120, ref. 128, *instead* of Du Seuil, Paris, 1970, *read* Seuil, Paris, 1970.

page 121, insert reference 164a:

164a W. M. Fitch, *Nature*, 229 (1971) 245.

COMPREHENSIVE BIOCHEMISTRY

Section I — Physico-Chemical and Organic Aspects of Biochemistry
Volume 1. Atomic and molecular structure
Volume 2. Organic and physical chemistry
Volume 3. Methods for the study of molecules
Volume 4. Separation methods

Section II — Chemistry of Biological Compounds
Volume 5. Carbohydrates
Volume 6. Lipids — Amino acids and related compounds
Volume 7. Proteins (Part 1)
Volume 8. Proteins (Part 2) and nucleic acids
Volume 9. Pyrrole pigments, isoprenoid compounds, phenolic plant constituents
Volume 10. Sterols, bile acids and steroids
Volume 11. Water-soluble vitamins, hormones, antibiotics

Section III — Biochemical Reaction Mechanisms
Volume 12. Enzymes — general considerations
Volume 13. (third edition). Enzyme nomenclature (1972)
Volume 14. Biological oxidations
Volume 15. Group-transfer reactions
Volume 16. Hydrolytic reactions; cobamide and biotin coenzymes

Section IV — Metabolism
Volume 17. Carbohydrate metabolism
Volume 18. Lipid metabolism
Volume 19. Metabolism of amino acids, proteins, purines, and pyrimidines
Volume 20. Metabolism of cyclic compounds
Volume 21. Metabolism of vitamins and trace elements

Volume 29, Part A

COMPARATIVE BIOCHEMISTRY, MOLECULAR EVOLUTION

Chapter III

Biochemical Evolution in Bacteria

J. DE LEY AND K. KERSTERS

Laboratory for Microbiology and Microbial Genetics, Faculty of Sciences, State University, Ledeganckstraat 35, 9000-Gent (Belgium)

I. INTRODUCTION

Micro-organisms, in particular bacteria and bacteria-like organisms, are almost certainly the oldest inhabitants of the earth. There is good evidence that rod-shaped bacteria lived here as long as roughly 3 billion years ago[186]. It seems likely that micro-organisms originated some 3.5–4 billion years ago. If we assume an average generation time of a few hours, then it appears that presently living bacteria have over a trillion generations behind them. The number of mutations than may have been inherited, stored, and now find expression is beyond imagination. By way of comparison we may recall that modern man is a few hundred thousand generations away from his ape-like ancestor the *Proconsul*, and only some three thousand generations from Neanderthal man. The variations in size and morphology of bacteria are very moderate. Even an experienced microbiologist is quite often unable microscopically to distinguish one genus from another. However, the variations in biochemical mechanisms surpass those in every other type of living being. Bacteria have undergone a very pronounced *biochemical* evolution and very little morphological evolution. Nearly all the other microscopic organisms show us the opposite, with a quite pronounced variety of often very beautiful morphological patterns. Those are the algae, the protozoa, the yeasts and the fungi.

II. ORIGIN OF BACTERIA

Let us first delineate the time span in which bacterial evolution occurred.

References p. 72

[1]

The earth originated about 4,5–5 billion years ago. There is a consensus of opinion that the early precambrian paleo-atmosphere contained no or only minute traces of oxygen. It was a tremendous chemical reaction vat[42] in which a great variety of organic compounds were formed. This material served a double purpose. Firstly, it contained the building blocks for the assembly of the primitive living beings. This process took about one billion years. It appears fairly sure that as long as some 3 billion years ago organisms existed that would be microscopically indistinguishable from present-day bacteria[186]. Secondly, the rest of this chemically synthesized material was used as a source of energy to permit continued life, growth and evolution of the primitive organisms.

It is certainly not our purpose to summarize all views on events that may have happened so long ago — with no one to witness them. Current thinking on this subject has been reviewed by Cloud[45], De Ley[57], Sagan[182] and Margulis[143]. Although opinions occasionally differ somewhat on details, the general trends of thought are very similar.

Events may be summarized briefly, as far as required for the following discussion. The first groups of living beings were anoxygenic and procaryotic. One of us[57] reasoned on biochemical grounds that chemiautotrophic organisms would be only of temporary advantage. In these organisms, the energy required for the incorporation of CO_2 in the Calvin cycle is provided by the oxidation of inorganic substances. In the early precambrian world the oxidant was obviously not O_2 but some inorganic reducible substance such as sulphate or carbonate. There is geochemical evidence that sulphate existed at that time. A considerable improvement arose when new organisms evolved, the anaerobic photoautotrophic bacteria, containing a pigment system to trap light energy in ATP. The next evolutionary step was the development of photosystem II, using water as proton and electron donor.

The concomitant O_2 liberation was going to change the face of the earth. Geochemical evidence indicates that a trace of O_2 was present in the atmosphere some 2 billion years ago. In the next 1.4 billion years it swelled to about 1% of the present concentration. During this period fundamental changes occurred. New organisms arose, with cytochromes and oxidative phosphorylation, using O_2 as terminal electron acceptor.

An enormous array of small eucaryotes developed. An attractive theory suggesting their formation by symbiosis was formulated by Margulis[143]. Her hypothesis was that anaerobic procaryotic organisms ingested (1) aerobic organisms, to be used as protomitochondria, (2) blue-green algae which were to function as chloroplasts, and (3) free-living spirochaete-like organisms to serve as the $(9+2)$ basal bodies of flagella. Classical mitosis is supposed to have evolved in these composite cells, millions of years after the evolution of photosynthesis.

This hypothesis led to extensive biochemical and molecular-biological research on the origin of mitochondria and chloroplasts[46, 47]. There appears to be a reasonable similarity between chloroplasts and blue-green algae[47]. Chloroplast rRNA hybridizes with DNA of blue-green algae. Plastid DNA codes for very many plastid components. These organelles may contain procaryotic pathways. The origin of mitochondria is less clear[47]. There are some points of similarity between mitochondria and aerobic bacteria, notably their size, the occurrence of naked DNA, of RNA, ribosomes, DNA and RNA polymerases, and formylmethioninyl tRNA initiation. However, mitochondria are biochemically as far removed from eucaryotes as from presently living bacteria. Their DNA-size is about 2% or less of the common bacterial genome, or about as large as a small bacteriophage genome. Mitochondria do not have mucopeptide or diamino-pimelic acid. More than 85% of mitochondrial protein is made by the eucaryotic nucleus; and even most of the mitochondrial ribosomal proteins are of nuclear origin. Not a single protein coded by mitochondrial DNA has been rigorously identified. In yeast, at least 63 nuclear genes affect mitochondrial function. There are many differences from the eucaryotic cell as well. Mitochondrial DNA does not hybridize with eucaryotic cell DNA. There is a new asymmetric mechanism of DNA synthesis in mito-chondria. The ribosomes are very small at 55 S. There is only one rRNA cistron in *Neurospora* mitochondrial DNA; the DNA-dependent RNA polymerase is different from both the eucaryotic and the bacterial enzyme, mitochondrial tRNA is unique, etc. If mitochondria are really of bacterial origin they have come to be so far removed from their source that it is frequently difficult to associate them with bacteria. Therefore a number of biologists prefer a monophyletic pattern, with a more or less continuous development of eucaryotic cells from procaryotes. The antagonism between both approaches suggests a variety of interesting experiments[47].

It has been realized by many biologists that one of the very important happenings in evolution was the separation of eucaryotes from procaryotes. New evidence was recently added from cytochrome c and t-RNA sequences[101, 149]. The evolutionary divergency between pro- and eu-caryotes is about 2.6 times as great as the divergencies between the eucaryotic kingdoms. The animals, green plants and fungi developed from a common ancestral stock, very different from bacteria and blue-green algae. It is assumed that the separation of the eucaryotic kingdom from the procaryotes occurred about one–two billion years ago. The assembly of the primitive eucaryotes started early, perhaps shortly after the formation of the blue-green protoalgae.

III. EVOLUTION OF BACTERIA

Some biologists tend to believe that precambrian bacteria were the same or nearly the same as presently existing ones; thus it is sometimes assumed that the primitive anaerobic heterotrophs were clostridia, that the primitive anaerobic bacterial photoautotrophs closely resembled the presently existing *Chlorobacteriaceae, Athio-* and *Thio-rhodaceae,* etc. This implies that presently living bacteria have changed little since the time of their origin, that they are some kind of living fossils, or at least the result of a bradytelic evolution.

Molecular-biological data, however, do not support this view[57]. DNA nucleotide sequences are quite different amongst genera. DNA of most bacterial genera do not hybridize or only very weakly. These differences can also be seen at the level of the individual cistrons. The amino acid sequences of cytochrome *c* are very different[52]. Base sequences of tRNA for the same amino acid in different taxa differ[101] by up to 58%. Ribosomal RNA cistrons of widely different bacterial genera (*e.g.* Gram-positive *versus* Gram-negative, or high %GC *versus* low %GC) have low DNA similarities[60]. Most bacterial genera are the product of a tremendous evolutionary divergency. There is no reason to believe that this evolution stopped long ago and that the ancient bacteria came to us unchanged. It is much more likely that a great variety of bacteria arose and disappeared with changing conditions, climate[27], available food, flora, fauna, etc. For example, it is unlikely that bacteria pathogenic for warm-blooded animals existed before these animals themselves originated some 225 million years ago. Likewise, it is very probable that phytopathogenic bacteria for, say, the angiosperms originated from some other ancestor less than 150 million years ago. Conversely, it is very likely that such extinct groups as the cordaites, the trilobites, the graptolites, etc. were plagued by their own disease-provoking bacteria, which no longer exist.

Some evolutionary theories have been proposed on a morphological basis. Stanier and Van Niel[198], suggesting that bacterial evolution went from simple to complex, hypothesized a primitive coccus type. Separate lines were thought to run through the micrococci, the polarly flagellated rods, the Gram-positive bacteria and actinomycetes, the peritrichously flagellated rods and the *Chroococcales*. According to Bisset[25] *Spirillum*, and, according to Leifson[132], aquatic polarly flagellated bacteria would be the primitive ones from which all others were derived. In view of modern data these hypotheses are now largely of historical interest, although some basic points, such as the divergency between Gram-positive and Gram-negative bacteria, remain meaningful.

Both Knight[123] and Lwoff[139] looked more at the physiological side of bacterial evolution. According to the latter author, the present-day hetero-trophic bacteria were derived by loss of CO_2 fixation from the autotrophs. In this fashion bacteria would arise with active carbohydrate metabolism, using ammonia as the main source of nitrogen. Representative species at this nutritional level extend from the N_2-fixing bacteria to the pseudo-monads and some *Enterobacteriaceae*. The free-living N_2-fixing bacteria such as *Azotobacter* etc. were thought to be extreme and more recent developments. It was suggested that the symbiotic nitrogen fixers, such as the rhizobia, evolved from the free-living forms. Further changes were thought to be the result of still more evolutionary losses, *e.g.*, there are a number of strains (*Salmonella*, *Shigella*, *Proteus* and *Vibrio cholerae*) which cannot use ammonia but require amino acids. These "exacting" strains are supposed to have originated by loss mutations from the non-exacting ones. Further losses in synthetic ability led to organisms which needed growth factors and vitamins. There are many examples of these: many clostridia, staphylococci etc. The origin of pathogenic organisms was likewise explained by loss of nutritional abilities. The authors mention that bacteria which lost the ability to synthesize complex molecules preferred to live in higher organisms. Life in such an environment might then offer the opportunity for the development of other properties, expressed as pathogenicity. The ultimate restriction of synthetic ability finally led to strict parasitism, where the necessary growth requirements can be supplied only by living tissues. An example is *Neisseria gonorrhoeae*.

This physiological theory may be summarized as follows. Bacteria in a primitive world acquired a complete array of enzymes. Such wholly self-sufficient organisms existed at a very early stage and the rest of evolutionary history consisted for them mainly in the loss of some properties, the most evolved progeny being the one with the fewest enzyme systems. Although certainly loss of properties might occasionally be useful — or at least not harmful — its occurrence is improbable as a general rule in the last 2 or 3 billion years. In some cases — for example, where the origin of symbiosis, pathogenicity and parasitism is concerned — we are inclined to agree with part of this hypothesis; yet, given the results of present-day research we do not think that the hypothesis of regressive nutritional evolution is generally tenable.

IV. THE RELATIVE TIME-SCALE

It is noteworthy that students of bacterial evolution do not have at their disposal the most important factor: a time-scale. For the Metaphyta and Metazoa, time was petrified in their fossils. Palaeontology, stratigraphy, geology, physical dating, comparative anatomy, embryology, etc. are important foundations to unravel evolution and phylogeny. On this evolutionary tree of ever-changing forms, the biochemical pathways of each group of organisms can, if known, be grafted and their most likely interrelation and evolution inferred.

Such a basis is not available for bacteria. The tree of bacterial evolution is largely concealed in a fog of ignorance extending from its roots some $4 \cdot 10^9$ years ago to the present terminal branches. The lack of an absolute time-scale is an enormous problem, as it prevents us from knowing the intricate branching in bacterial evolution. Fortunately, there exist at least three ways to tackle this problem. True, none of them provides an absolute time-scale in terms of years or aeons. But they can be used to construct a relative one enabling us to decide whether two or more organisms arose from the same ancestor or whether a particular organism is younger or older than another. For two of these criteria, part of the previous history of the cell has been frozen in some of its components. One is the nucleotide sequence in DNA or RNA, the other is the amino acid sequence of homologous proteins. The third approach compares various aspects of bacterial metabolism.

It was calculated, for both tRNA and cytochrome c[101] that the fixation rate is about $2 \cdot 10^{-10}$ actual replacements per nucleotide site per year.

1. DNA relatedness

Organisms which arose from a common ancestor may still have part of their DNA nucleotide sequences in common. There exist methods to measure quantitatively this degree of relatedness. The correct meaning of DNA hybridization for taxonomic and phylogenetic interpretation should be briefly clarified. It is currently the fashion to work at T_m *minus* $25°C$. In these conditions DNA:DNA hybrids can be formed only between heterologous stretches with less than 20–25% mispairing[64]. Above this value there is apparently very little probability that sufficient nucleation centers exist. Two heterologous DNAs may have considerable base alignment homology but a rather small number of unsuitably positioned mispairings may prevent hybridization. Therefore, when two bacterial DNAs do not hybridize at T_m *minus* $25°C$, they may still be related. Lack of DNA hybridization has no taxonomic or phylogenetic meaning. Lower temperatures should be more thoroughly explored. The higher the experimental DNA hybridization values, the more meaningful the actual relationship between both organisms involved. Except for very high levels ($> 80\%$), the degree of DNA hybridization is probably always lower than the base alignment homology, and lower also than phenotypic percent similarity. Remote relationships can be explored by study of the individual cistrons such as cytochrome *c*, or DNA:rRNA hybridizations[60]. Still, the results of DNA–DNA and DNA–rRNA hybridizations represent a good approximation of the phylogenetic tree of closely related organisms. The role of molecular biology in the elucidation of bacterial phylogeny has been reviewed recently[57]. The degree of DNA relatedness is not always in harmony with the orthodox classification of bacteria. For example, strains within one genetic race of *A. tumefaciens* have a DNA relatedness of at least 80%; different genetic races of the same species share less than 50% DNA relatedness, but the genera *Escherichia* and *Shigella* stand at 85% DNA relatedness. On the whole, however, the difference between two genera is much greater. Quite generally speaking, the evolutionary divergency in bacteria, as measured by DNA relatedness, is enormous. This is also apparent from the enormous variation in biochemical pathways and known amino acid sequences of bacterial cytochrome *c*.

This field is being investigated thoroughly, but it is still too early to present a complete relatedness-tree for the microbial world. For the moment we know only of a few small branches here and there as only a limited number of hybridizations have been carried out, *e.g.* in and amongst

Pseudomonas and *Xanthomonas*[63], *Agrobacterium* and *Rhizobium*[62, 95], *Neisseria*[122], *Enterobacteriaceae* (see below) and some other taxa. In most cases it is not yet possible to relate these data to the biochemistry of the organisms concerned, because too little is known about them. In the few cases where this is possible, the conclusions are very instructive. We shall discuss only two examples.

(a) The Enterobacteriaceae

One case concerns the *Enterobacteriaceae*. Both physiologically and biochemically they are reasonably well known. Their enzymic carbohydrate pathways were briefly reviewed by one of us[56].

Their DNA relatedness can be calculated from the results obtained by Brenner *et al.*[29–32]. Fig. 1 is only a simplified summary of the complex relationships.

Two genera, *Escherichia* and *Shigella*, appear to be very closely related, some 85% of their chromosome being still nearly perfectly identical. Within the part they have in common, about 1% of the bases differ. The similarity is

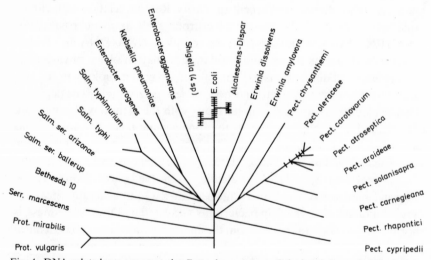

Fig. 1. DNA relatedness amongst the *Enterobacteriaceae*. Calculated from the results obtained by Brenner *et al.*[30,32]. The rod lengths indicate degrees of DNA relatedness. The short perpendicular lines in a few species (*e.g.* *E. coli*) represent the strain differences within one species. Not all published data could be represented in the drawing. Temperature used: 60–66° in 0.28 *M* phosphate buffer.

so great that one can argue that the shigellae can be included as one or more species within the genus *Escherichia*. It appears likely that the separation of both genera is a reasonably recent phenomenon in the history of this family. Both genera occur in the intestines, and mainly in the intestines of primates. It seems quite conceivable that the pathogenic *Shigella* arose from the non-pathogenic *Escherichia* through loss of flagella, of the formic hydrogen-lyase enzyme system and of a few other biochemical features, as well as through acquisition of new antigenic structures and of toxin production. There are no indications as to when *Shigella* arose, but dysentery was a common disease in antiquity since both Greek and Roman writers refer to it. It is quite possible that *Shigella* arose somewhere on the evolutionary road from ape to man, during the last 25 million years.

The alcalescens-dispar bacteria are about 90% DNA homologous with *E. coli*. They are as closely related to *E. coli* as are most *E. coli* strains. It was therefore proposed[30] that the A–D strains should definitely be included as part of the genus *Escherichia*. A similar situation occurs in *Pectobacterium*, *P. oleraceae*, *P. atroseptica*, *P. aroideae* and *P. solanisapra* are over 85% DNA homologous with *P. carotovorum*. Strains of *P. carotovorum* themselves are at least 80% related with the reference strain. The border cases in this family are *Serr. marcescens*, and *Proteus* with less than 12% DNA homology. Not all genera are homogeneous. Many strains of *Erwinia* and *Pectobacterium* share only about one fourth or one third of their genome[32,85].

Most of the *Enterobacteriaceae* are derived from a common origin, but diversified considerably. Very few genera share more than one-third of their chromosome, the rest of the genome being already so different that DNA–DNA hybridization does not occur. The thermal stability of the intergeneric DNA hybrids is generally low, showing that nucleotide sequences no longer match perfectly and that mutational changes have already deeply affected the common intergeneric DNA part. This divergency underlies the differences in fermentation patterns and other phenotypic features between the genera.

(b) Agrobacterium

Another case, very thoroughly understood both genotypically and phenotypically, concerns the agrobacteria[58,62,64,119]. Most of these bacteria are plant pathogens. Many strains, known as *Agrobacterium tumefaciens*, provoke tumours in a disease called "crown gall". *A. rhizogenes* causes in addition an excessive root outgrowth, known as "hairy root disease". Some strains are not pathogenic and are called *A. radiobacter*. There are a few other species whose position is doubtful[59, 95] and they will not be considered here.

Over 250 strains were investigated. DNA hybridizations, the determination of the thermal stability of the DNA hybrids, the determination of over a hundred different physiological and biochemical features of living cells, and of the electrophoretic patterns of soluble proteins (see below IV. 3.b, p. 67), show that these bacteria form three main and eight minor genetic races. The full results will be published elsewhere.

Fig. 2 represents their genetic relationship as known from DNA hybridizations. The genetic races are extremely uneven in size. Two small ones consist of just one crown-gall strain each, one from Iran and the other from South Africa, different from each other and from all the other strains. The third race consists of three crown-gall strains from the U.S.A. which all need growth factors. The fourth race (cluster 2) contains some 53 strains. Here are all the hairy root disease, rhizogenic organisms and the atypical crown gall organisms.

All the other races form the large cluster 1 with 150 strains. It consists of two large and five small races. Within each race the genomes are at least 80% similar, and the DNA hybrids are very thermostable. We call one large subgroup B6 and the other one TT111, for representative strains. All the typical tumefaciens strains belong here.

The explanation of all our results is that the ancestral agrobacteria split long ago first into five genetic races. Since gymnosperms are hosts for crown-gall (M. de Cleene, to be published), we suppose that the division occurred less than 300 million years ago. We do not know whether it happened at once (unlikely) or in phases. It probably started with mutations in the third base of the triplets, followed by mutations in the second and first bases. This second step occurred mainly outside the active center of the enzymes. In this fashion a heterogeneous population arose, very much like the present cluster 1. The third step consisted of more mutations, now in the active centers as well, eliminating some enzymes and creating others. The five groups diversified to their present level of some 20% DNA homology and 75% phenotypic similarity. Only cluster 1 and cluster 2 expanded in number. Cluster 2 remained very homogeneous.

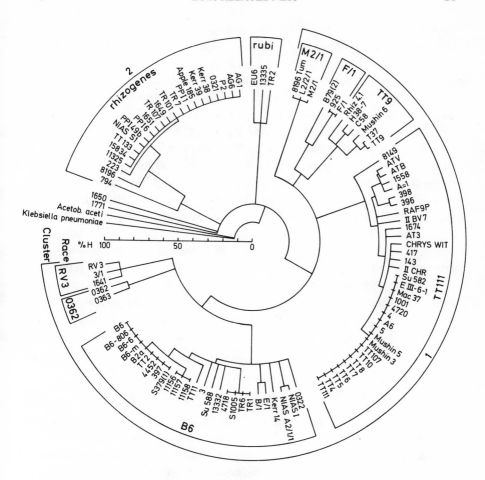

Fig. 2. DNA relatedness within the genus *Agrobacterium*. The symbols represent strain numbers. Figure modified from ref. 58.

In spite of the fact that strains have been isolated in the U.S.A., Japan, Australia and several European countries, they are all remarkably alike. The nucleotide sequences in their genomes, their phenotypic features and protein patterns are nearly indistinguishable.

In cluster 1 history repeated itself. As a result of the mutational steps one and two, described above, cluster 1 itself became heterogeneous, and was split into 7 races. Only the races B6 and TT111 gave rise to large populations. The former appears to occur mainly around the Pacific Ocean

(U.S.A., Japan, Australia), the latter around the Atlantic Ocean (U.S.A., Canada, Europe, S. Africa). Of the other races, fewer than five strains are known. All seven races have about 50% DNA homology, but phenotypically they are over 88% similar. Within each race the DNA homology is over 80% and the phenotypic differences concern a few features. It seems to us that these seven races will in the future diverge more and more, thus giving rise to phenotypically new species.

Our results together with those of Brenner et al. (loc. cit.), show very clearly that a nomenclatorial species may vary considerably from one genus to another. In Agrobacterium, a species extends over a span of 80% DNA relatedness (e.g., A. tumefaciens B6, 1771, 1650 and 1651 share only 20% DNA relatedness). In the Enterobacteriaceae it is quite different. Twenty-seven strains of Escherichia coli display more than 70% DNA relatedness amongst themselves[30]. In the few instances known, the genetic differences amongst Salmonella strains are small. Brenner et al.[29] found a DNA related-ness of about 90% between Salm. typhimurium and Salm. typhi. In our labora-tory we established that there is even less difference between Salm. typhimu-rium and Salm. pullorum (De Ley, Reynaerts and Cattoir, unpublished).

2. Evolution of metabolic processes

(a) Theory of metabolic pathway evolution

Horowitz[103] was the first to propose that biosynthetic pathways grow by a process of retrograde evolution. Primitive organisms in the primitive ocean used a number of preformed organic compounds. Upon exhaustion of each of them, those mutants were favoured which had evolved an enzyme (-system) capable of producing the exhausted compound from its precursor(s).

The new genes presumably originated by gene duplication, one of them mutating to produce the new enzyme[133]. Genes of a metabolic pathway, in sequence in the chromosome, were supposed to have a common evolutionary origin and genes for the enzymes of a pathway were supposed to have arisen from one another[104].

Hegeman and Rosenberg[96] reviewed experimental advances in gain muta-tion. So far, four mechanisms are known by which bacteria acquire the ability to metabolize new compounds and by which a metabolic pathway can grow: (1) an inducible enzyme becomes constitutive; (2) the specificity of an enzyme changes; (3) the sensitivity to adverse metabolites produced from a novel compound decreases; (4) permeability is acquired for a previously impermeable compound.

The examples of acquisitive evolution, however, suggest an alternative to Horowitz's theory[104]: metabolic pathways could develop by mutation of a variety of genes, not necessarily of adjacent ones. A useless or not so useful cistron can thus apparently be transformed into one of immediate necessity. It has been suggested that cistrons under a single control were assembled in an operon by translocation[96]. Gene duplication might have been important at very early times, when genomes were still rather small.

Regulation of metabolic pathways by feedback inhibition was probably a later development.

(b) Catabolism of carbohydrates (Fig. 3)

There are two main catabolic processes for carbohydrates which are very widely distributed in the living world: glycolysis and the hexosemonophosphate oxidative cycle (shunt). As far as is known, both systems occur in most, if not all, metaphyta and metazoa and in many — but not all — bacteria. It seems extremely likely that both pathways are very old and that at least one of them occurred in bacteria which were the ancestors of the presently living world. Photo- and chemi-autotrophic organisms use a variation of the shunt to incorporate CO_2 as glyceraldehyde phosphate (Calvin cycle), and part of the glycolytic pathway to produce hexoses. From the present viewpoint heterotrophs came first, which would make the shunt the ancestor of the Calvin cycle.

A completely active glycolytic pathway occurs in the Enterobacteriaceae, the homolactic acid bacteria, the saccharolytic clostridia and probably in the propionic acid bacteria and some bacilli. Many other heterotrophic bacteria possess a glycolytic pathway which lacks one or at most two enzymes. Thus the heterolactic acid bacteria and bifidobacteria lack fructose-1,6-diphosphate aldolase; pseudomonads, acetic acid bacteria and Hydrogenomonas saccharophila usually lack either fructose-6-phosphate kinase or the above-mentioned aldolase. Obviously, these are all cases of regressive evolution.

The shunt is likewise present in many bacteria, such as the Enterobacteriaceae, the saccharolytic clostridia, the homofermentative lactic acid bacteria, the pseudomonads, acetic acid bacteria and certainly in many other

Organisms	Many aerobic Gram-negative bacteria in the range 50-70% GC	Hydrogenomonas saccharophila	Acetic acid bacteria: Acetobacter and Gluconobacter	Pseudomonas	Enterobacteriaceae: Salmonella, Escherichia, Shigella, Aerobacter, E. freundii	Bacillus	Saccharolytic butyric acid bacteria: saccharolytic butanol–aceton bacteria; saccharolytic butanol–isopropanol bacteria; Cl. kluyverii Clostridium
Remarks				Great variations in substrate decomposition: carbohydrates, amino acids, aromatic compounds, hydrocarbons	Lactic and acetic acids, butanediol, propylene glycol	Great variations in fermentation types	
Some specializations in metabolism		Fac. H_2 autotrophic	Many Some enzymes on cytoplasmic membrane		Variations in pyruvate breakdown		
Cytochromes	+	+	+	+	+	+	−
Entner–Doudoroff pathway	+	+	− and +	+	+	−	− (one strain + ?)
Shunt	probably +	incomplete	+	+	+	probably +	+ and −
Glycolysis		incomplete	incomplete	incomplete	+	+	+ and −

Fig. 3. Schematic survey of the occurrence of the main mechanisms of carbohydrate breakdown in bacteria.

aerobic bacteria as well. Here too regressive enzyme losses occurred. They are, however, more diversified than in the glycolysis. *E.g. Hydrogenomonas saccharophila* lacks gluconate-6-phosphate dehydrogenase, *Bifidobacterium* lacks glucose-6-phosphate dehydrogenase and the heterofermentative lactic acid bacteria lack the transketolase–transaldolase system.

In organisms in which both glycolysis and the shunt are defective, another mechanism on the same level took over. In the case of the hetero-fermentative lactic acid bacteria and bifidobacteria, it is the phosphoketolase system, and for *H. saccharophila*, it is the Entner–Doudoroff system. It seems very likely that each of these systems arose independently in the ancestors of these bacteria before both the glycolytic pathway and the shunt lost an enzyme by mutation. Defects coupled with the absence of an auxiliary pathway resulted in lethal mutants. As far as we know, the phosphoketolase

minolytic	Purinolytic	Butyri-bacterium	Homofer-menters	Hetero-fermenters	Bifido-bacterium			Myxobacteria ↑ Cytophaga	Actinomycetes ↑ Mycobacterium ↑ Corynebacterium
						Succinate bacteria ↑ Propionibacterium			
				Lactic acid bacteria					
Protein and mino acid reakdown	Purine and pyrimidine breakdown	Lactate breakdown	Phosphoketolase pathways			Methylmalonyl pathway			
−	−		usually −	−	−	+		+	+
−	−	− (1 strain?)	−	−	−	−		−	−
		+	No trans-ketolase and transaldolase	No G6P dehydro-genase	+				
+ and −		+	No aldolase	No aldo-lase	+				

systems are confined to the lactic acid bacteria.

Decker *et al.*[53] pointed out that both lactic and propionic acid bacteria assume a very special position. The former produce catalase; the latter possess cytochromes, catalase and the Krebs cycle. These enzymes are features of aerobic bacteria. It was therefore suggested that both groups of bacteria arose by regressive evolution from aerobic bacteria through extensive losses in the cytochrome system. It is possible that these changes were accompanied by the development of the phosphoketolase and methylmalonyl pathways.

One of us[55,56] suggested that organisms with the Entner–Doudoroff scheme are closely related phylogenetically and constitute a separate branch of the evolutionary tree. This pathway consists of the following sequence of reactions:

glucose→glucose 6-phosphate→gluconate 6-phosphate→2-keto-3-deoxygluconate 6-phosphate→pyruvate + glyceraldehyde 3-phosphate

References p. 72

Recent work undertaken to establish the occurrence of this pathway[61,118] confirmed the above hypothesis. Some 270 strains, belonging to about 40 different genera were examined. The pathway was detected in some 25 genera. It is very striking that nearly all of them are Gram-negative rods with a % GC in the range of 50–70. The Entner–Doudoroff pathway is absent in all Gram-positive organisms investigated, except perhaps in *Streptococcus faecalis*[195] and in a few *Nocardia* strains. It occurs in the following genera: *Pseudomonas, Xanthomonas, Flavobacterium, Chromobacterium, Protaminobacter, Azotobacter, Azomonas, Microcyclus, Aeromonas, Acetobacter xylinum, Gluconobacter, Rhizobium, Mycoplana, Agrobacterium sensu stricto (tumefaciens, radiobacter, rhizogenes), Agarbacterium, Zymomonas, Hydrogenomonas, Rhodopseudomonas, Pasteurella,* the family of the *Enterobacteriaceae* (except *Enterobacter agglomerans*). We enumerate all these genera because they are probably on the same phylogenetic branch. It is interesting to note that most of these organisms are aerobic; they are either polarly or peritrichously flagellated. The facultative anaerobic *Zymomonas* is very likely a later development. In the *Enterobacteriaceae*, the E.D. pathway is only present when induced on gluconate; it is thus still a genetic possibility and suggests a remote relationship between this family and the *Pseudomonadaceae.* "*Pseudomonas*" *rubescens* 519 (perhaps an *Alteromonas*) is a very remarkable organism. With its low % GC of 46 and low hybridization of DNA it is genetically very different from the pseudomonads *sensu stricto*[63]. Because of the presence of the Entner–Doudoroff pathway we imagine that its ancestor once belonged to *Pseudomonas* and diverged considerably, except in cell shape, Gram stain, flagellation and carbohydrate metabolism. This may also be true of the low % GC, polarly flagellated marine *Alteromonas*[20].

A related type of metabolism was discovered[61] in Gram-negative, aerobic, peritrichous bacteria closely related to *Alcaligenes denitrificans* with 64–70% GC. The pathway is

$$\text{D-gluconate} \rightarrow \text{2-keto-3-deoxy-D-gluconate} \xrightarrow{\text{ATP}} \text{2-keto-3-deoxy-D-gluconate 6-phosphate} \rightarrow$$
$$\text{pyruvate} + \text{glyceraldehyde 3-phosphate}$$

Dehydration precedes phosphorylation. These organisms apparently constitute a side-branch of the previous group. It is possible that some *Erwinia*s have a similar mechanism[118]. The meaning of the occurrence of a gluconate dehydrase in *Clostridium aceticum*[12] is not clear.

There exist a large number of strictly anaerobic bacteria (see Prévot *et al.*[172]). The best known, biochemically speaking, are the clostridia, the anaerobic chemi-autotrophic sulphate-reducing *Desulfovibrio* and carbonate-reducing *Methanobacteriaceae* organisms, as well as the photo-autotrophic *Chlorobacteriaceae* and *Thiorhodaceae*. Decker *et al.*[53] hold that the clostridia are still closest to the ancestral anaerobic bacteria. One of us[57] pointed out that presently living bacteria are already some 3 or 4 billion years removed from their ancestors; many changes have occurred.

(c) Catabolism of aromatic compounds by Pseudomonads

Pseudomonads can be subdivided into a number of subgeneric clusters[197]. A small one, termed the acidovorans group, made up of the species *Pseudomonas acidovorans* and *P. testosteroni*, metabolize aromatic compounds in a special fashion, summarized in Fig. 4.

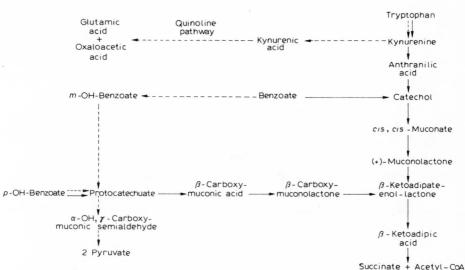

Fig. 4. Survey of aromatic catabolism by fluorescent (——) and acidovorans (- - - -) pseudomonads.

In most pseudomonads the decomposition pathways of tryptophan, benzoate and *p*-OH-benzoate converge at β-keto-adipate-enol-lactone, to end up finally as succinate and acetyl-CoA. The acidovorans organisms have different mechanisms: tryptophan is broken down by the quinoline pathway, benzoate and *p*-OH-benzoate by way of protocatechuate to pyruvate. It seems very likely that the acidovorans organisms represent a separate phylogenetic branch.

(d) Energy conservation

In order to grow and to live, organisms require energy for the biosynthesis of their polymeric constituents (proteins, polysaccharides, lipids, polynucleotides), many other complex organic molecules (vitamins, co-enzymes, etc.), and for complex processes (osmotic regulations, movement, etc.). Except in some photochemical reactions where light energy is trapped as chemical energy, energy is moved from molecule to molecule until it reaches its final destination. Whenever it leaves its material carriers, it is frequently dispersed as heat and lost forever.

Although many compounds carry energy around in the cell, it is very striking that ATP is the only general carrier. The remarkable fact that it is a nucleotide undoubtedly has some evolutionary meaning which is not completely understood. It may be recalled that the entire machinery for protein synthesis appears to be very similar in all organisms. Therefore the conclusion seems inescapable that ATP is a very early energy-transporting substance needed for the biosynthesis of proteins; both mechanisms existed several billion years ago in the ancestors of all modern cells. The role of energy carriers in primitive life will be discussed by one of us elsewhere (De Ley, Proc. Roy. Soc., London, to be published).

It was only afterwards that evolutionary divergency into different cell types occurred. As this happened very early, long before morphological divergency, the first mutations occasioned biochemically diverse organisms. The results are still with us, with present-day bacteria showing little morphological diversity but an immense variety of biochemical pathways. Most of these are catabolic mechanisms of substrate decomposition and different ways to produce ATP. This is mainly the subject of the comparative carbohydrate metabolism of bacteria as summarized by one of us[56].

ATP is produced in a variety of reactions of which only a small number has been unravelled (Fig. 5). The real molecular mechanism of ATP production in photosynthesis of micro-organisms and of plants, in the electron transport mechanism (*e.g.* the cytochrome systems) of aerobic organisms, and in the oxidations of various inorganic substrates (H_2, NH_3, NO_2^-, etc.) is not known. The mechanism of ATP production is as yet

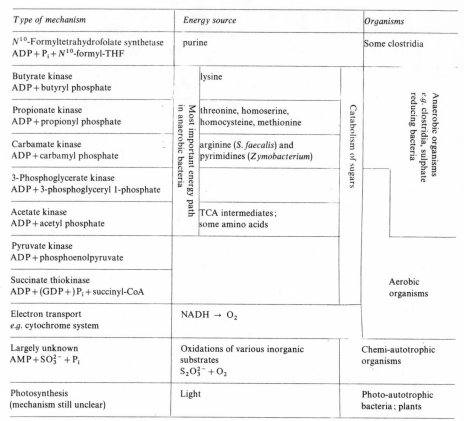

Type of mechanism	Energy source			Organisms
N^{10}-Formyltetrahydrofolate synthetase $ADP + P_i + N^{10}$-formyl-THF	purine			Some clostridia
Butyrate kinase $ADP + $ butyryl phosphate		lysine		
Propionate kinase $ADP + $ propionyl phosphate		threonine, homoserine, homocysteine, methionine		
Carbamate kinase $ADP + $ carbamyl phosphate		arginine (*S. faecalis*) and pyrimidines (*Zymobacterium*)		
3-Phosphoglycerate kinase $ADP + $ 3-phosphoglyceryl 1-phosphate				
Acetate kinase $ADP + $ acetyl phosphate		TCA intermediates; some amino acids		
Pyruvate kinase $ADP + $ phosphoenolpyruvate				
Succinate thiokinase $ADP + (GDP+)P_i + $ succinyl-CoA				Aerobic organisms
Electron transport *e.g.* cytochrome system	$NADH \rightarrow O_2$			
Largely unknown $AMP + SO_3^{2-} + P_i$	Oxidations of various inorganic substrates $S_2O_3^{2-} + O_2$			Chemi-autotrophic organisms
Photosynthesis (mechanism still unclear)	Light			Photo-autotrophic bacteria; plants

column label: Most important energy path in anaerobic bacteria

column label: Catabolism of sugars

column label: Anaerobic organisms *e.g.* clostridia, sulphate reducing bacteria

Fig. 5. Some mechanisms of ATP production by bacteria. Partially from Decker *et al.*[53].

known only at the substrate level for a number of heterotrophic bacteria. The following reactions are near-ubiquitous in both aerobic and anaerobic organisms:

3-phosphoglycerate kinase
$$ADP + \text{3-phosphoglyceryl-1-phosphate} \rightarrow ATP + \text{3-phosphoglyceric acid}$$

pyruvate kinase
$$ADP + \text{phospho-enolpyruvate} \rightarrow ATP + \text{pyruvate}$$

acetate kinase
$$ADP + \text{acetyl phosphate} \rightarrow ATP + \text{acetic acid}$$

The former two enzymes occur as part of the glycolytic pathway in most organisms. In bacteria with an incomplete glycolysis (see above), glyceraldehyde 3-phosphate is provided not only by aldolase, but also by either the shunt + phosphoketolase or by Entner–Doudoroff path-way. The importance of both enzymes in the autotrophic bacteria is perhaps negligible.

The third enzyme, acetate kinase, is probably important mostly in anaerobically growing bacteria, producing acetic acid as one end-product of fermentation, *e.g.* the *Enterobacteriaceae*, the clostridia, the lactic acid bacteria, propionic acid bacteria etc. In aerobic bacteria, acetyl-CoA or acetyl phosphate is metabolized *via* the Krebs cycle, and the ATP gain may be as much as twelve-fold.

Succinate thiokinase from higher organisms catalyzes the reactions

$$\text{Succinyl-CoA} + \text{Pi} + \text{GDP} \rightarrow \text{succinate} + \text{CoA} + \text{GTP}$$

and ATP is produced by nucleosidediphosphate kinase

$$\text{GTP} + \text{ADP} \rightarrow \text{GDP} + \text{ATP}$$

In bacteria the reaction is somewhat simpler with succinyl-CoA synthetase

$$\text{Succinyl-CoA} + \text{Pi} + \text{ADP} \rightarrow \text{succinate} + \text{ATP}$$

Although the distribution of this enzyme is not yet well known, it seems very likely that it is, in one form or another, specific for aerobic bacteria using the Krebs cycle; it, too, may therefore be a later acquisition.

A number of very specialized ATP-generating systems are known, such as the carbamate kinase, propionate kinase, butyrate kinase and the N^{10}-formyl THF synthetase (see Fig. 5). Most of them occur in anaerobic bacteria, such as clostridia and *Zymobacterium*, and it is therefore not impossible that the ancestors of these enzymes existed as long as 2 billion years ago. It is as yet impossible to assess the proper position and significance of other ATP-generating enzymes in the energy economy of the micro-world and the evolution of energy-providing pathways.

(e) Control mechanisms

(i) Allosteric regulation of DAHP synthase

Related micro-organisms seem to possess identical allosteric control characteristics of branch-point enzymes. The comparative study of the allosteric pattern of control for 3-deoxy-D-arabino-heptulosonate-7-phos-phate (DAHP) synthase and citrate synthase are two well-documented examples, which will be discussed briefly.

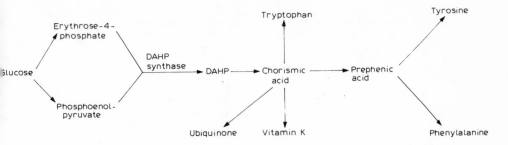

Fig. 6. Overall picture of pathways for the formation of aromatic amino acids and vitamins in *E. coli*.

DAHP synthase catalyzes the initial biochemical reaction of the multi-branched pathway for the biosynthesis of aromatic amino acids in micro-organisms (Fig. 6). Chorismic acid is a precursor of three aromatic amino acids, of ubiquinone, vitamin K and folic acid. A balanced biosynthesis of the aromatic amino acids can be achieved through appropriate inhibition of branch-point enzymes by the end-products.

Jensen *et al.*[110] surveyed the pattern of allosteric regulation of DAHP synthase in 32 microbial genera represented by more than 90 species. No fewer than seven alternative patterns of allosteric regulation of DAHP synthase were detected, indicating that significantly different control mechanisms for a single enzyme have developed during evolution. The different types of feedback inhibitions are summarized in Table I. Explanation of unusual terminology and a review of the pathways of bio-synthesis of aromatic amino acids and their control can be found else-where[89].

(1) Sequential feedback inhibition of DAHP synthase occurs in *Bacillus*, *Sporosarcina ureae*[111], *Staphylococcus*, *Gaffkya*, *Flavobacterium*, *Achromo-bacter parvulus* and *Alcaligenes viscolactis*. DAHP synthase is inhibited by intermediates (chorismic acid and prephenic acid) and not by the end-products phenylalanine, tryptophan and tyrosine. All investigated strains of *Bacillus alvei* were anomalous in that their DAPH synthase was competitively inhibited by tryptophan. *B. alvei* differs also from all other *Bacillus* species in its low % GC (32.5%) and in some biochemical features.

(2) Isoenzymatic feedback inhibition is displayed by all enteric bacteria investigated (*Escherichia*, *Aerobacter*, *Salmonella*, *Shigella*, *Serratia* and

TABLE I

Patterns of feedback inhibition of bacterial DAHP synthases[110, 111]

Type of feedback inhibition	Occurrence
a. Sequential feedback inhibition	Bacillus
	Sporosarcina ureae
	Staphylococcus
	Gaffkya tetragena
	Flavobacterium devorans
	Achromobacter parvulus
	Alcaligenes viscolactis
b. Isoenzymatic feedback inhibition	
dominant phe-sensitive isoenzyme	Escherichia
	Erwinia carotovora
dominant tyr-sensitive isoenzyme	Aerobacter
	Serratia
	Erwinia amylovora
	Aeromonas
mainly phe- and tyr-sensitive isoenzymes	Salmonella
	Shigella
	Saccharomyces cerevisiae
	Neurospora crassa
c. Concerted or multivalent feedback inhibition	non-sulphur purple bacteria
d. Cumulative feedback inhibition	Hydrogenomonas
e. DAHP synthase inhibited by tyrosine	Pseudomonas
	Neisseria
	Mycobacterium, Nocardia
f. DAHP synthase inhibited by phenylalanine	Alcaligenes faecalis
	Achromobacter viscosus
	Veillonella alcalescens
g. DAHP synthase inhibited by tryptophan	Myxococcus
	Streptomyces
	Micromonospora

Erwinia) and *Aeromonas*. A balanced control is possible because an inhibitable isoenzyme exists for each major end-product. Feedback inhibition of DAHP synthase isoenzymes was also demonstrated in *Saccharomyces cerevisiae* and *Neurospora crassa*.

(3) Cumulative feedback inhibition of DAHP synthase was present in *Hydrogenomonas* strains.

(4) Concerted or multivalent feedback inhibition of DAHP synthase was found in photosynthetic non-sulphur purple bacteria. The simultaneous

presence of the three end-products is required before any significant in-hibition of DAHP synthase can occur.

(5) DAHP synthase inhibited by tyrosine. This type of inhibition occurs in representatives of *Pseudomonas*, except *P. acidovorans* and *P. testosteroni.* The latter strains possess tyrosine and phenylalanine inhibitable DAHP synthases. Both inhibitors act together in a less than cumulative fashion. These differences in control pattern of *P. acidovorans* and *P. testosteroni* are in agreement with the known biochemical differences between the acidovorans group and the other aerobic pseudomonads[197]. *Xanthomonas* shows close evolutionary links with *Pseudomonas*[63] but displays totally different allosteric control mechanisms: chorismic acid, but not prephenic acid, is a powerful inhibitor for DAHP synthase of *X. hyacinthi* and *X. campestris,* suggesting a special type of sequential feedback inhibition.

(6) DAHP synthase inhibited by phenylalanine. DAHP synthase of *Alcaligenes faecalis* and *Achromobacter viscosus* are inhibited solely by phenylalanine, in contrast to the DAHP synthase of *Alcaligenes viscolactis* and *Achromobacter parvulus* which display a sequential type of feedback inhibition (see Table I). Indeed we know (De Ley and Shewan, un-published) that the latter two organisms are quite different from the former; *A. parvulus,* for example, is polarly flagellated.

(7) DAHP synthase inhibited by tryptophan was detected in *Myxococcus xanthus* and *several Streptomyces* and *Micromonospora* species.

The qualitative pattern of control exerted by aromatic amino acids upon DAHP synthase was almost without exception (B. alvei and Xanthomonas) a markedly conserved feature of related bacteria and seems to constitute a reliable generic characteristic. The impression is thus strengthened that genera are biological units, the results of evolutionary divergency, which conserved not only certain morphological, physiological and biochemical pathways but also the typical mechanisms to control their flow. Two taxa with the same type of control mechanism in the same pathway might — but need not — be related. When a taxon seems to be related to two other ones, the type of control mechanism may indicate its most probable phylogenetic neighbour. This procedure has been applied to *Sporosarcina ureae* and *Aeromonas formicans*[111]. The former has the characteristics of both *Bacilleae* and *Micrococcaceae*, but its control pattern links it to the spore formers. The latter has the characteristics of both *Enterobacteriaceae* and *Pseudomonas*, but its control pattern, which is similar to *Escherichia*, ties it to the enteric bacteria.

(ii) Allosteric regulation of citrate synthase

Citrate synthase (EN* 4.1.3.7) is a key enzyme of the tricarboxylic acid cycle and catalyzes the condensation of acetyl-CoA and oxaloacetic acid. A comparative study[222, 224] of the effects of the metabolic regulators NADH, AMP and α-ketoglutarate indicated that citrate synthases of approximately 50 bacterial strains fall into 3 distinct groups, correlating with present taxonomic divisions. The data of Table II are generalized, because not all citrate synthases were investigated for molecular weight[223] and inhibition by α-ketoglutarate[222].

The micro-organisms investigated could be divided into two main groups according to the molecular weight of their citrate synthase and the susceptibility of this enzyme to inhibition by NADH. There is a clear correlation

TABLE II

Patterns of allosteric regulation and approximate molecular weight of bacterial citrate synthase[222, 224]

Molecular weight > 250 000; inhibition by NADH. Gram-negative bacteria		Molecular weight ± 80 000; no inhibition by NADH and α-ketoglutarate. Gram-positive bacteria
Inhibition by α-ketoglutarate and no reactivation by AMP	Reactivation by AMP and no inhibition by α-ketoglutarate	
Escherichia[a,b]	Azotobacter[a]	Micrococcus
Proteus[a]	Pseudomonas[a,b]	Staphylococcus
Salmonella[a,b]	Xanthomonas	Corynebacterium
Klebsiella	Flavobacterium	Microbacterium
Aerobacter[a,b]	Chromobacterium	Bacillus[a]
Hafnia	Acinetobacter[a,b]	Arthrobacter
Erwinia	Moraxella	Brevibacterium
Serratia	Vibrio	Kurthia[a,b]
Pasteurella		Cellulomonas
		Mycobacterium[b]
		Nocardia
		Streptomyces[a,b]

[a] Inhibition by α-ketoglutarate[222].
[b] Molecular weight of citrate synthase was determined by gel chromatography[223].

* EN: *Enzyme Nomenclature*, 3rd edn., 1973, Amsterdam (Vol. 13 of this Treatise).

between molecular size of a citrate synthase, its allosteric regulatory behaviour and Gram stain. It is possible that citrate synthase is only susceptible to allosteric inhibition by NADH, when identical or similar subunits are specifically associated to a larger aggregate.

Gram-positive bacteria possess the smaller citrate synthase, displaying no regulatory mechanism. The possibility cannot be excluded that another type of allosteric control is operational in these micro-organisms. Because some enzymes are associated with the cytoplasmic membrane, differences in membrane structure between Gram-negative and Gram-positive micro-organisms may be correlated with the observed differences in allosteric control of citrate synthase.

The Gram-negative bacteria can be subdivided into two groups on the basis of response of the citrate synthase to AMP and α-ketoglutarate (see Table II). In one group, comprising strict aerobes such as *Pseudomonas*, *Azotobacter*, etc., the inhibition by NADH can be relieved by AMP whereas citrate synthase is not inhibited by α-ketoglutarate. In the other group (mostly *Enterobacteriaceae* or related organisms) AMP is without effect and citrate synthase is inhibited by α-ketoglutarate. The allosteric control mechanisms for citrate synthase emphasize the differences between the facultative anaerobic *Enterobacteriaceae* and related organisms (*Pasteurella*) on the one hand and the aerobic Gram-negative bacteria (*Pseudomonas, Xanthomonas, Acinetobacter, Azotobacter*, etc.) on the other. The inhibition of citrate synthase of *Enterobacteriaceae* by α-ketoglutarate can be understood as a typical case of end-product inhibition of the initial enzyme of a pathway[222]. There are indeed indications that even during the aerobic growth of *E. coli* on glucose the TCA cycle functions primarily for the biosynthesis of α-ketoglutarate.

(iii) Different induction patterns for the same overall metabolic pathway

Most pseudomonads decompose benzoic and *p*-OH-benzoic acids by way of the β-ketoadipate pathway (Fig. 4). This very widespread mechanism occurs also in *Acinetobacter, Mycobacterium, Nocardia, Bacillus*, etc. However, both the enzymes and the inducer mechanism are quite different[196]. The pathways were studied *in extenso* both in *Ps. putida* and in a strain of *Acinetobacter* (*"Moraxella calcoacetica"*). The nature of the inducers for each enzyme is quite different in both organisms. This shows that control mechanisms also evolve.

The real background has not yet been investigated but we suggest that it is the regulator and/or the operator gene which mutated so as to permit a different spectrum of inducers. One does not necessarily have to invoke a different evolutionary origin of the same pathway; evolution of the regulation mechanisms will do as well. Immunologically the picture is as follows. The enzymes for the decomposition of benzoic and p-OH-benzoic acids in *P. aeruginosa, P. putida, P. fluorescens* and *P. stutzeri* are immunologically related, but already distinguishable from one another. The enzymes of *P. multivorans* effect the same reactions but are not related immunologically: however, the control system is still the same. We believe this points to some mutations in the structural genes, very much like the structural differences in cytochrome c in many organisms. In the case of still more remote organisms such as *e.g. Acinetobacter*, the next step in evolutionary difference consisted of mutational changes in regulator and/or operator genes. But when mutation became very pronounced at all genic levels, a completely different pathway arose as was the case in the acidovorans group.

(f) CO_2 fixation in chemo- and photo-autotrophic bacteria

Autotrophs build up their cellular carbohydrates from CO_2. This reductive process requires 2 moles of ATP per mole of CO_2 incorporated. Both chemo- and photo-autotrophs appear to possess an analogous enzyme system for CO_2 incorporation. However, the mechanisms for ATP production and for reduction are quite different.

(i) Mechanism of CO_2 incorporation

There is a fair amount of evidence that the reductive (or photosynthetic or Calvin) pentose cycle exists in plants, algae and autotrophic bacteria alike. This cycle is too well known to require extensive discussion here. It must be a very old mechanism. It may have existed in more or less its present form as long as 3 billion years ago in precambrian autotrophic bacteria, and passed apparently almost unchanged through the blue-green and other algae to the plants. It has been detected in whole or in part in the chemoautotrophs *Thiobacillus, Hydrogenomonas, Nitrobacter, Nitrocystis* and in the photoautotrophs *Chromatium, Chlorobium, Rhodopseudomonas* and *Rhodospirillum*.

However, it seems possible that the green *Chlorobacteriaceae* incorporate CO_2 mainly by the alternative route of the reductive carboxylic acid cycle[80,192] (Fig. 7).

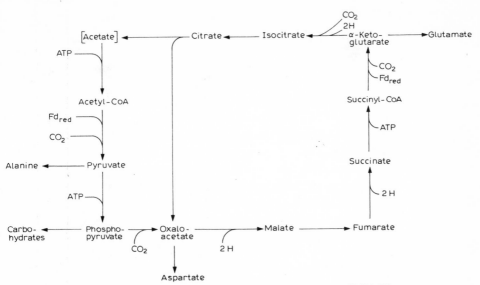

Fig. 7. The reductive carboxylic acid cycle in some *Chlorobacteriaceae*[80,192,194].

Chlorobium thiosulfatophilum appears to fix most of its CO_2 by two new ferredoxin-dependent reactions, catalyzed by pyruvate synthase and α-keto-glutarate synthase: acetyl-CoA + CO_2 + 2 Fd·e + 2 H$^+$ → pyruvate + CoA + 2 Fd, and succinyl-CoA + CO_2 + 2 Fd·e + 2 H$^+$ → α-ketoglutarate + CoA + 2 Fd. The most important intermediates in this cycle are α-ketoglutarate, oxaloacetate and pyruvate (required for the synthesis of the primary amino acids glutamate, aspartate and alanine), acetyl-CoA (required for the biosynthesis of lipids) and phosphoenolpyruvate (starting-point for the formation of carbohydrates). With the proper enzymic connections, this cycle can thus apparently help make the essential building blocks of the cell. The reductive pentose phosphate cycle is present but only weakly active. The occurrence of the reductive carboxylic acid cycle in other groups of organisms deserves further study. For the moment it appears that the *Chlorobacteriaceae* are in a very special position amongst the photo-autotrophs in that they possibly constitute a separate phylogenetic branch.

(ii) ATP production in photoautotrophs; photosystems I and II

The photosynthetic apparatus in bacteria is contained in small organelles called chromatophores or thylakoids. Usually they are vesicular, sometimes (*Rhodospirillum molischianum, Rhodomicrobium vannielii*) lamellar. The

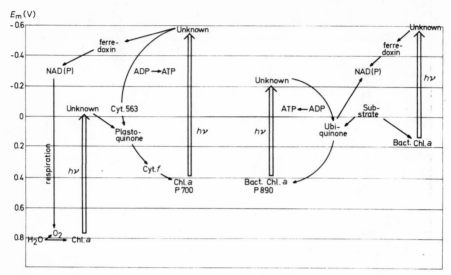

Fig. 8. Simplified comparison of plant (left) and bacterial (right) *(Chromatium, Rhodospirillum rubrum)* photosynthesis.

thylakoids of blue-green algae are likewise commonly lamellar. The chromatophores of the other algae and the chloroplasts of the higher plants have a much more complex structure (on their possible origin, see above).

The mechanism of photosynthesis in the blue-green algae is very similar to that in other algae and green plants. Of course they all produce O_2 by photosystem II which is the outstanding feature of green plant photosynthesis as well as the main difference from bacterial photosynthesis. Plant photosynthesis contains two photochemical steps connected in series (Fig. 8). In the most important one, system I, an electron is kicked up by light energy to a higher energy level. When it falls back through cyclic photophosphorylation, ATP is formed. If this electron is going to be used for the production of NADPH, another one is to be supplied by the photochemical reaction in system II. Thereby both ATP and O_2 are formed.

The exact mechanism of bacterial photosynthesis has not yet been determined. It, too, contains a cyclic photophosphorylation system for ATP production. The mechanism of formation of the reduced co-enzymes is not yet clear. One currently accepted theory assumes that the reducing agents (H_2S, thiosulphate, H_2, etc.) replace the electron in system I.

Chromatium and *Rhodospirillum rubrum* appear to contain a second photochemical reaction in parallel[98, 157, 202], where NADPH is produced by a photochemical reaction from the donor H_2A, independently from system I.

(g) Biosynthesis of tetrapyrroles (Fig. 9)

It is perhaps not insignificant that corrin derivatives (cobalamines, vitamin B_{12} and derivatives) appear to occur mainly in anaerobes such as in some clostridia (transformation of glutamate into methylaspartate by glutamate mutase in *Cl. tetanomorphum*), in methane bacteria (reduction of bicarbonate to methane), in propionic acid bacteria and in anaerobic photosynthetic bacteria[53]. It is very likely that corrin compounds existed

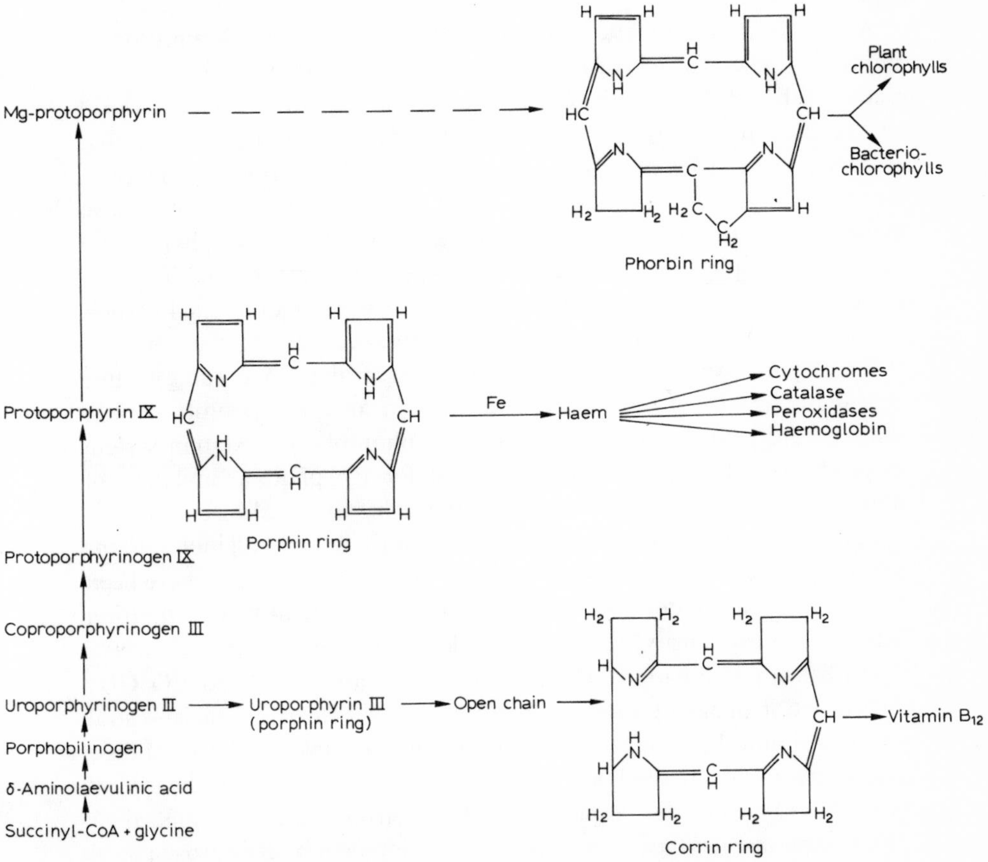

Fig. 9. Schematic pathway of tetrapyrrole biosynthesis.

References p. 72

very early in bacterial history, since they are still so widespread. They can be synthesized by ordinary heterotrophs, such as *E. coli* and *Streptomycetes*.

One can imagine that the primitive anaerobes later developed the enzymes for the synthesis of the porphin ring for cytochromes, catalase and peroxidases. Porphin is more oxidized than corrin and contains a methine bridge between the rings I and IV. These haem enzymes appear to be present in nearly all bacteria except in the strictly anaerobic clostridia and methane bacteria, and in some of the supposedly revertant lactic and propionic acid bacteria. Cytochromes are also present in the anaerobic photosynthetic bacteria; cytochrome c_3 is predominant in the sulphate-reducing *Desulfovibrio*.

A modified biosynthetic pathway leads by a way of Mg-protoporphyrin to phorbin derivatives, the chlorophylls, which are active in photosynthesis. Photosynthetic bacteria are unique in the bacterial world in that they have both porphins and phorbins. It is quite possible that phorbins arose from porphins when an ancestral anaerobic chemi-autotroph evolved the first photo-autotrophs[57]. The various chlorophylls have rather similar structures (Fig. 10). The tetrapyrrole of chlorophyll *a* and bacteriochlorophyll *a* respectively differs only in two groups and bacteriochlorophyll *c* differs from the other chlorophylls in not more than 4 groups.

Olson[165] proposed an interesting hypothesis concerning the common origin of photosynthetic bacteria and blue-green algae. According to this hypothesis, more than 3 billion years ago an ancestral photobacterium utilized light with chlorophyll *a* in a cyclic photophosphorylation system to produce ATP. The latter was required for the photo-assimilation of some organic compounds. The cycle supposedly consisted of chlorophyll *a*, a primary electron acceptor, a cytochrome and a quinone, all built into the cell membrane. Neurosporene or lycopene carotenoids may also have been present. Subsequent development of thylakoids and pigment concentration led to greater cell density (all below water level) and competition for the blue light (440 nm blue band of chlorophyll *a*). Mutation at R_1 to $CO \cdot CH_3$ and reduction in one pyrrole produced bacteriochlorophyll in the ancestral purple sulphur bacteria, now absorbing in the violet and the orange. Thereby, however, the redox potential decreased below that of the system H_2O/O_2; O_2 could not be produced, the electron had to be supplied by other reactions, and the full evolutionary development of their progeny was forever strangled. Only a few groups of photosynthetic micro-organisms

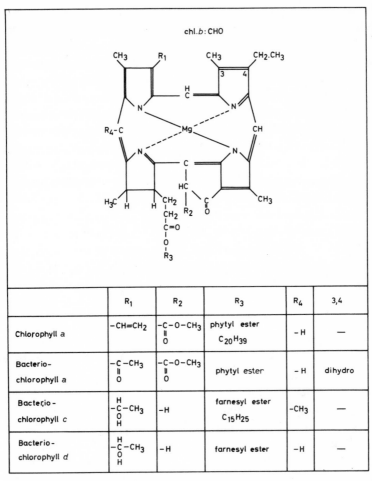

Fig. 10. Similarities and differences between chlorophylls of plants and bacteria, respectively.

	R_1	R_2	R_3	R_4	3,4
Chlorophyll a	$-CH=CH_2$	$-C-O-CH_3$ $\parallel$ O	phytyl ester $C_{20}H_{39}$	$-H$	—
Bacterio- chlorophyll a	$-C-CH_3$ $\parallel$ O	$-C-O-CH_3$ $\parallel$ O	phytyl ester	$-H$	dihydro
Bacterio- chlorophyll c	$\begin{smallmatrix}H\\-C-CH_3\\O\\H\end{smallmatrix}$	$-H$	farnesyl ester $C_{15}H_{25}$	$-CH_3$	—
Bacterio- chlorophyll d	$\begin{smallmatrix}H\\-C-CH_3\\O\\H\end{smallmatrix}$	$-H$	farnesyl ester	$-H$	—

arose, evolving into the present-day *Thiorhodaceae*, *Athiorhodaceae* and *Chlorobacteriaceae* (a separate side-line, see above). Their pigments absorb mainly in the far-red and red region above 700 nm, and in the yellow to violet range (400–600 nm). These organisms neither were nor are very much hampered by shading by algae and plants. This was brought about by a few changes in the ancestral bacteriochlorophyll, mainly at the positions R_1, R_2, R_3 and R_4.

References p. 72

The exact phylogenetic position of the latter organisms is not clear. They are all Gram-negative, suggesting an evolved, rather complex cell wall. The *Athiorhodaceae* have a cytochrome oxidase system and are aerobic in the dark. Did they evolve from the other photosynthetic bacteria by mutational gain of cytochrome oxidase? Or, conversely, did the *Chlorobacteriaceae* and *Thiorhodaceae* arise from the non-sulphur purple bacteria by a number of steps also involving the loss of cytochrome oxidase? All these bacteria are now so many generations away from their ancestors that their phylogeny is difficult to trace.

A noncyclic electron-transport chain required for the photo-assimilation of CO_2, arose independently. In the chlorophyll *a* containing organisms, the proton and electron donor was water. This line led to ancestral blue-green algae and later on to all O_2-producing photosynthetic organisms.

(h) Gram character and cell-hull structure

Generally speaking, bacteria may be divided rather neatly into a Gram-positive and a Gram-negative group. It is now quite clear that this is due to fundamental differences in the cell-wall structures.

The cell wall of Gram-positive organisms consists mainly of one layer of a giant sac-like peptidoglycan (mucopeptide or murein) onto which teichoic acids and polysaccharides are attached. The cell wall of Gram-negative bacteria is much more complex. The peptidoglycan layer is thinner and is surrounded by at least 2, and frequently 3, layers of proteins, polysaccharides and lipoproteins. Although there are many clustered variations the general pattern is the same in many bacteria.

One cannot avoid the impression that a fundamental split in the biosynthesis of cell walls occurred very early in the history of bacteria, such that the Gram-positive and Gram-negative bacteria represented two distinct phylogenetic branches. Later on further ramifications led to changes in the peptidoglycans of the cell-wall and in the phosphatides of the membrane.

(i) Cell-wall peptidoglycans

Many variations occur in the amino-acid compositions and primary structures of cell-wall peptidoglycans from different micro-organisms. In most bacterial walls so far examined, the glycan portion of the peptidoglycan layer consists of alternating β-1,4-linked *N*-acetylglucosamine and *N*-acetylmuramic acid residues. The average chain length varies from 10 to 65 disaccharide units[87].

(ii) The peptides

The peptide moiety consists of two or three parts. It contains two equal tetra-or penta-peptide subunits, and frequently an amino acid or peptide bridge linking both subunits together. A number of lactate carboxyl groups of the *N*-acetylmuramic acid residues form amide linkages to the terminal L-alanine of the subunit. By way of example, the primary structure of the peptidoglycan from *Micrococcus roseus*[170] is shown in Fig. 11. The structure of the bacterial peptidoglycans has been reviewed[87,88,185]. We shall here refer mainly to the recent summary presented by Kandler's school in München[185]. The peptide moiety varies considerably from one taxon to another. The peptide subunit consists of L-alanine (usually) → D-glutamic acid (usually) → an. L-diamino acid (usually) → D-alanine → D-alanine (not always).

Most variations occur in the interpeptide bridge and in the mode of cross-linkage. There are two main groups called A and B. Group A connects the ω-amino group of the diamino acid 3 of one subunit to the carboxyl group of D-alanine 4 of the other subunit. The linkage may be either direct (subgroup A1) or through a variety of peptide bridges (subgroups A2, A3 and A4). The B group connects the α-carboxyl group of D-glutamic acid 2 of one subunit to the carboxyl group of D-alanine 4 of the other subunit. Several kinds of bridges are known.

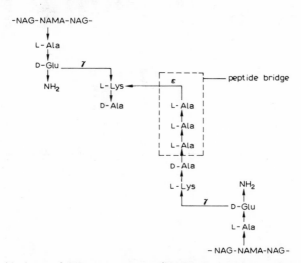

Fig. 11. Peptidoglycan of *Micrococcus roseus* (NAG = *N*-acetylglucosamine; NAMA = *N*-acetylmuramic acid).

References p. 72

TABLE III. Some examples of peptide subunits and peptide bridges in bacterial peptidoglycans[185]

Micro-organism	Type	Peptide subunit	Peptide bridge
Staphylococcus aureus	A3α	L-Ala–D-Glu(NH$_2$)–L-Lys–D-Ala	L-Lys–Gly$_{5-6}$
Staphylococcus epidermis	A3α	see above	L-Lys–Gly$_2$–L-Ser–Gly$_2$
Micrococcus luteus, flavus, etc.	A2	L-Ala–D-Glu(Gly)–L-Lys–D-Ala	L-Lys–D-Ala–L-Lys–D-Glu(Gly)–L-Ala
Micrococcus roseus, lactis, etc.	A3α	L-Ala–D-Glu(NH$_2$)–L-Lys–D-Ala	L-Lys–L-Ala$_{3-4}$
Planococcus	A4α	L-Ala–D-Glu–L-Lys–D-Ala	L-Lys–D-Glu
Sporosarcina ureae	A4α	see above	L-Lys–Gly–D-Glu
Sarcina maxima and ventriculi	A3γ	L-Ala–D-Glu(NH$_2$)–LL, Dpm–D-Ala	LL, Dpm–Gly
Aerococcus	A1α	L-Ala–D-Glu(NH$_2$)–L-Lys–(D-Ala)	direct 3–4
Gaffkya homari	A1α	see above	direct 3–4
Streptococcus / Peptostreptococcus	A1α, A3α and A4α	see above	twelve different types
Thermobacterium	A4α	see above	L-Lys–D-isoAspN
Streptobacterium casei, etc.	A4α	see above	see above
Sporolactobacillus plantarum and inulinus	A1γ	L-Ala–D-Glu(NH$_2$)–m-Dpm(NH$_2$)–(D-Ala)	m-Dpm direct
Betabacterium	Mostly A4α and A4β	L-Ala–D-Glu(NH$_2$)–L-Orn–D-Ala or L-Ala–D-Glu(NH$_2$)–L-Lys–D-Ala	frequently L-Lys–D-Asp or L-Orn–D-Asp
Leuconostoc	A3α	see above	several L-Lys–L-Ser–L-Ala types
Bifidobacterium	A3α, A3β, A4α, A4β	see above	several types
Coryneforms	different A and B types	various types	great variety
Cellulomonas	A4β	L-Ala–D-Glu–L-Orn–D-Ala	L-Orn–D-Asp; L-Orn–D-Glu
Arthrobacter	A3γ	see above	L,L-Dpm–Gly$_{1 \text{ and } 3}$
	A3α		L-Lys–L-Ala$_{1-4}$; L-Lys–L-Thr–L-Ala$_{1-3}$, etc.
Brevibacterium	A1γ	see above	m-Dpm-direct
Corynebacterium sensu stricto	A1γ	see above	see above
Nocardia; Mycobacterium	A1γ	see above	m-Dpm-direct
Propionibacterium and aerobic actinomycetes	A3γ	see above	L,L-Dpm–Gly
	A1γ		m-Dpm-direct
Clostridium	mostly A1γ	see above	m-Dpm-direct
Bacillus	mostly A1γ	see above	see above
Gram-negative bacteria	probably mostly A1γ	see above	see above

Several examples of peptide moieties are given in Table III. The structures alone cannot be used for taxonomic or phylogenetic considerations because the same type of peptidoglycan frequently occurs in widely different genera. For example, the Alγ type occurs in many Gram-negative bacteria, *Bacillus*, many corynebacteria, etc.; the A3 and A4 types are very widespread. However, on the whole there is a good correlation between the peptidoglycan type and several phenotypic and genotypic features within most Gram-positive bacteria, thus making it a good taxonomic characteristic.

Schleifer and Kandler[185] proposed a possible phylogenetic relationship between peptidoglycan subgroups and variations. They suggested that the multilayered peptidoglycans of the A3 and A4 groups represent a primitive stage. Most of the Gram-positive genera belong here, *i.e.* nearly all *Micrococcaceae* and *Lactobacilleae: Staphylococcus, Micrococcus, Sarcina maxima* and *S. ventriculi, Streptococcus, Peptostreptococcus, Thermobacterium, Streptobacterium, Betabacterium, Leuconostoc, Pediococcus, Bifidobacterium* as well as *Arthrobacter, Cellulomonas* and various coryneforms, some propionibacteria, some aerobic actinomycetes (*e.g. Streptomyces*, etc.). These cell-wall types are considered more primitive because they are chemically complex, variable, multilayered and the result of multiple gene action.

These organisms diversified considerably in % GC, in phenotypic features, and, to a limited extent, in their peptidoglycan structure. Six other cell-wall types arose either from the previous types or from a common ancestor.

(*1*) Some micrococci have a multilayered peptidoglycan with one or more peptide units as bridge. They are *M. luteus*, some other closely related micrococci and aerobic *Sarcina* strains.

(*2*) *Aerococcus* and the related *Gaffkya homari* have a multilayered peptidoglycan with direct cross-linkage through Lys–Ala.

(*3*) A number of coryneforms and the anaerobic actinomycetes have a cell wall with very special cross-linkages between positions 2 and 4. The unusual type suggests that these organisms have an unusual metabolism.

(*4*) Many corynebacteria, aerobic actinomycetes, *Mycobacterium*, most of the aerobic and micro-aerophilic *Bacillus*, and the anaerobic *Clostridium* have a multilayered peptidoglycan with the directly cross-linked Alγ type. It is known that there are close biochemical and physiological links between *Bacillus* and *Clostridium*[56] on the one hand, and between many coryneforms, *Mycobacterium* and the actinomycetes on the other. A possible link between both groups is a new working hypothesis.

(5) Most, if not all Gram-negative bacteria have monolayered peptidoglycans with an m-Dpm direct 3–4 cross-linked A1γ type. The rigidity of the cell wall is here partially provided by the additional layers of lipoproteins, proteins, etc. It is interesting to recall[56] the similarities in carbohydrate metabolism in the saccharolytic clostridia — *Bacillus* — *Enterobacteriaceae* — *Pseudomonas* series. In the *Myxobacteriales* only patches of the A1γ peptidoglycan remain, allowing the flexibility of the cells.

(6) The *Spirochaetales* display a minor variation on the above theme, with L-Orn.

(iii) Bacterial phosphatides

Phosphatides (Fig. 12) from bacteria are largely localized in the cytoplasmic membrane. The lipid content of membranes from Gram-positive bacteria varies from 13 to 40%. These bacteria have practically no lipids in the cell wall. The cell walls of Gram-negative bacteria contain lipoproteins and lipopolysaccharides, but apparently no phosphatides.

A review[107] of the occurrence of these compounds in a great variety of bacteria shows that phosphatidic acid, phosphatidyl glycerol and diphosphatidyl glycerol (cardiolipin) occur in most bacteria, both Gram-positive and Gram-negative ones. It seems plausible to assume that these three phosphatides are of ancient origin and occurred even in primitive bacteria. Other phosphatides occur in a more limited number of taxa.

Phosphatidyl ethanolamine appears to be widespread in Gram-negative bacteria, in the Gram-positive *Bacillus*, perhaps in *Clostridium*, and irregularly in *Mycobacterium*. The non-sporeforming Gram-positive *Micrococcaceae*, *Lactobacillaceae*, *Propionibacteriaceae*, *Corynebacteriaceae*, and *Actinomycetales* are believed, on morphological and biochemical grounds, to be closely related phylogenetically. Phosphatidyl ethanolamine is almost entirely absent from these taxa, a circumstance which appears to strengthen the relationship. Because of similarities in the fermentation patterns, the genus *Bacillus* is supposed to be distantly related to the *Enterobacteriaceae*[56]. This relationship is again illustrated by the common occurrence of phosphatidyl ethanolamine.

When NH_3^+ is substituted by $N^+(CH_3)_3$, phosphatidyl choline or lecithin results. This is the principal phosphatide of both higher plants and animals. In bacteria it occurs mainly in the photo-autotrophs. Phosphatidyl serine and phosphatidyl inositol occur rather unevenly in various genera, although

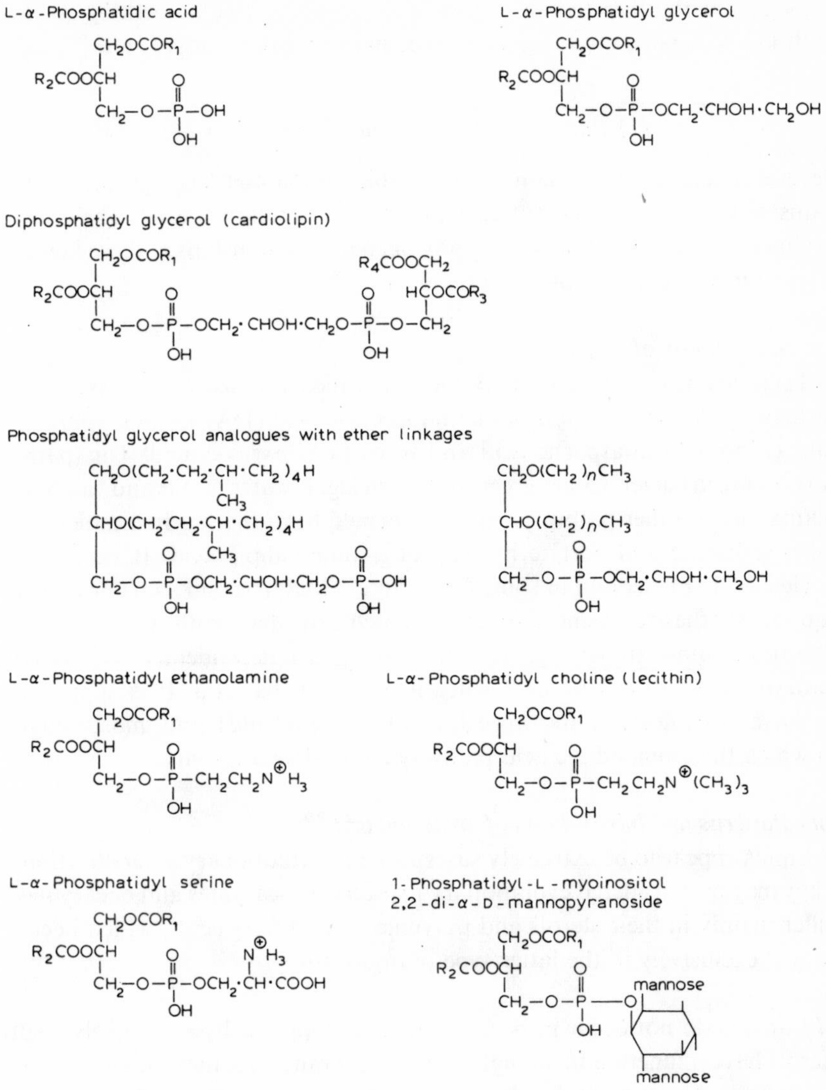

Fig. 12. Phosphatides in bacteria.

the former compound appears to be present in some pseudomonads, in a number of *Enterobacteriaceae* and in *Bacillus*, whilst the latter is fairly widespread in *Mycobacterium* as a myo-inositol mannoside. Some very special cases are known. *Halobacterium* contains some phosphatidyl glycerol analogues with ether linkages; they may possibly have something to do with ion transport, given the halophilic nature of these organisms.

(i) Pathways of more general occurrence

Several biochemical mechanisms are stable within one large group of organisms such as bacteria, but vary at a higher taxonomic level. They may eventually be useful in unravelling phylogenetic relationships at these higher levels. Only a few examples will be given.

(i) Biosynthesis of lysine[220]

There are two main pathways for lysine biosynthesis. In one, lysine is formed by decarboxylation of diaminopimelic acid (DAP),which contains four carbons from aspartic acid and three from pyruvic acid. This pathway occurs in bacteria, blue-green and green algae, water molds and vascular plants. In the other pathway, lysine is formed from the succinyl moiety of α-ketoglutarate and acetate by way of α-aminoadipic acid. It occurs in euglenids, in higher and in some lower fungi. Metazoa and some protozoa cannot synthesize lysine. Vogel[220] concluded that both patterns are genetically quite stable and that they emerged independently. The DAP path appears to be the oldest, since it occurs in bacteria. It is supposed to have been lost in the ancestors of the euglenids and most fungi, in which the aminoadipic acid pathway evolved as a novelty.

(ii) Patterns and biosynthesis of some lipids[26, 79]

Lipids appear to be extremely susceptible to evolutionary diversification. They may be considered evolutionary markers. Procaryotes and eucaryotes differ mainly in their sterols and polyunsaturated fatty acids, which occur almost exclusively in the latter type of organism.

(1) Sterols do not occur in bacteria and cyanophyta. It seems likely that sterols have a function in strengthening membrane structures in eucaryotic organisms. They do not possess a mucopeptide in the cell wall and Bloch[26] suggested that the evolutionary loss of the mucopeptide was partially offset by increased sterol incorporation in the membrane.

(2) Unsaturated fatty acids. Mono-unsaturated fatty acid can be synthesized in a variety of ways. Photosynthetic and some eubacteria follow an

anaerobic pathway, exemplified by

$$C_8 \xrightarrow{C_2} \Delta^3\text{-}C_{10} \xrightarrow{4C_2} \Delta^{11}\text{-}C_{18}$$

A second mechanism desaturates fatty acids *in situ* by oxidation according to

$$C_{18} \xrightarrow{O_2} \Delta^9\text{-}C_{18}$$

It appears to be very widespread. It occurs in several bacteria and actinomycetes, in cyanophyta, all protista investigated, in red algae, fungi and vertebrates. A third pathway seems plausible, but remains to be clarified in detail. It occurs in higher plants, green algae and some euglenids.

There also appears to be a number of quite distinct lines in the distribution of polyunsaturated fatty acids. Linoleic acid can be desaturated either at C_{15} with the formation of α-linolenic acid or at C_6 for γ-linolenic acid, followed by arachidonic acid. Bacteria do not produce polyunsaturated fatty acids; cyanophyta, green algae and higher plants form

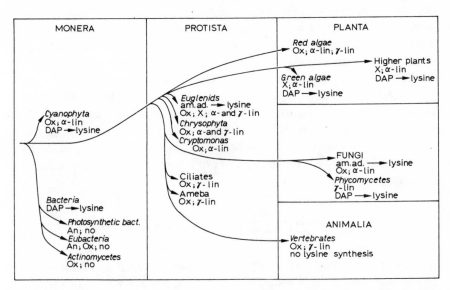

Fig. 13. The occurrence of various pathways for the metabolism of unsaturated fatty acids and for the biosynthesis of lysine in the living world. The data from Vogel[220] and Bloch[26,79] are projected onto the five-kingdom system described by Whittaker[225]. Abbreviations: synthesis of mono-unsaturated fatty acids by the anaerobic (An), the oxidative (Ox) and the plant (X) pathways. Desaturation of linoleic acid with the formation of either α-linolenic (α-lin) or γ-linolenic (γ-lin) acids. Absence of linoleic acid desaturation (no). Biosynthesis of lysine from either diaminopimelic acid (DAP) of from α-amino-adipic acid (am.ad.).

α-linolenic acid. The protista have three possibilities: some produce either one of the linolenic acids separately, others make both. Both pathways occur in fungi. In vertebrates the biosynthetic chain is incomplete because they cannot convert oleate to linoleate, though the latter can still be oxidized to γ-linolenate. In Fig. 13 we projected the summary of unsaturated fatty acid formation on the five-kingdom systems described by Whittaker[225].

3. Relationship of proteins

The sequence of amino acids in a protein is a translated copy of the sequence of the purine and pyrimidine bases in the corresponding gene. Mutations in the nucleotide codons of a gene, directing the synthesis of a given protein, may be expressed by amino acid substitutions in the primary structure of the protein. By comparing the molecular structure of the gene products (proteins) we can unravel their evolutionary relationships and trace their ancestry. Proteins possessing a pervading similarity in their amino acid sequences are structurally homologous and related[52] in an evolutionary sense.

The evolutionary history of mitochondrial cytochrome c is now well-documented. Because the rate of evolution of this protein seems to have been relatively constant (1% change per 20 million years), there is hope that cytochrome c can be used as a "paleontological clock" to date the divergence of eucaryotic organisms from a common evolutionary pathway. In the following section we wish to outline the progress that has been made in the molecular approach to the study of evolutionary relationships between bacterial proteins.

(a) Structure and amino acid sequences of low molecular weight proteins

(i) Cytochrome c

Cytochrome c seems to be a protein eminently suited to evolutionary biochemical studies. It is widespread in nature, occurring in most aerobic and in many anaerobic micro-organisms. Most cytochromes c are readily purified and possess a small molecular weight; they are thus well-suited for sequencing by standard methods. Their rate of mutation seems to be optimal for phylogenetic studies of widely diverse organisms including animals, plants and micro-organisms[52].

Studies on sequence differences[82] and on the reconstructed ancestral

sequences[52] of more than 45 different mitochondrial cytochromes c made
possible the design of phylogenetic trees with the same overall topology
as phylogenetic trees derived from morphological, paleontological or other
data[52,82,142,150,161]. Considerable homology still exists in primary
structures of cytochrome c from widely divergent taxa such as yeasts and
mammals.

It is evident that similar studies on the primary structures of bacterial
cytochromes c will throw light on the phylogenetic relationships between
bacterial cytochromes c in particular and on the evolution of bacteria in
general. However, there exist so many different types of bacterial cytochromes c and so few primary structures are known that construction of
significant phylogenetic trees is as yet out of the question. Only 13
complete sequences of bacterial cytochrome c are known at present but
this number is expected to increase exponentially in the near future.

The structural[19,113], functional[102] and evolutionary[5,126] aspects of
bacterial cytochromes c have been reviewed.

(1) Distribution and classes of bacterial c-type cytochromes

The microbial biochemist is faced with an enormous variety of bacterial
c-type cytochromes. Comparative biochemistry indicates the existence of at
least 10 classes of cytochromes where the prosthetic group (haem) is
covalently bound to the protein: c_2, c_3, c_4, c_5, c_7, c', cc', flavin-c, c-555
and several monohaem c-type cytochromes of low molecular weight. This
subdivision is at present only tentative and will probably change in the
future when more properties and primary structures of these proteins are
known. The various groups of c-type cytochromes differ in molecular weight,
spectral characteristics, redox potential, isoelectric point and number of
haem groups. Their exact functions are not always known. The situation is
even more complex because one bacterial strain can possess more than one
c-type cytochrome. This multiplicity has been demonstrated in some photosynthetic bacteria, *Desulfovibrio vulgaris*[35], *Azotobacter vinelandii*[200],
Pseudomonas denitrificans[108], and *Bacillus subtilis*[154,155].

Cytochromes c are ubiquitous among micro-organisms, with the exception of strictly anaerobic fermenters such as the clostridia. They occur
in strict aerobes such as *Azotobacter* and *Pseudomonas*, as well as in
facultative anaerobes (denitrifiers), in photosynthetic (*e.g. Chromatium,
Rhodospirillum, Chlorobium*) and chemosynthetic bacteria (*e.g.* the strictly
anaerobic sulphate-reducing *Desulfovibrio* and the ammonia- and nitrite-

oxidizers *Nitrosomonas* and *Nitrobacter*).

Photosynthetic bacteria may be divided into three categories — green sulphur, purple sulphur and purple non-sulphur — because of nutritional and morphological characteristics[151], and the distribution of different c-type cytochromes. The *Chlorobacteriaceae* possess a typical c-555 cytochrome. They lack c_2 and c' cytochromes. The *Thiorhodaceae* (*Chromatium*) have no c_2 and no c-555 but do possess cc' (RHP) and a flavin-containing cytochrome c. The *Athiorhodaceae* (e.g. *Rhodopseudomonas* and *Rhodospirillum*, except *Rhodopseudomonas gelatinosa*) characteristically possess c_2 and c'- or cc'-type cytochromes, but no c-555 and flavin cytochromes.

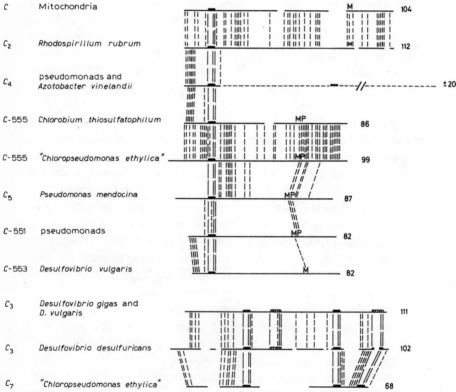

Fig. 14. Sequence similarities of several bacterial cytochromes c. Sequences are aligned on the basis of identical positions of haem-binding cysteines. ■■■, –Cys–A–B–Cys–His–haem-binding sites; ▥▥, putative –Cys–A–B–C–D–Cys–His– haem-binding sites; ––––, regions not yet sequenced; M, sixth haem ligand residue methionine; MP, methionine–proline sequence. Important deletions are indicated by gaps. The total number of amino acids per molecule is indicated at the end of the chain. Adopted from Van Beeumen[216].

(2) Primary structures of bacterial cytochromes c. One can predict from the great differences in physico-chemical properties that unlike the mitochondrial cytochromes c, procaryotic cytochrome c has undergone a tremendous evolution. Indeed, the available sequence data reflect as much (see above). The question then arises whether all these cytochromes c are to be considered homologous in the evolutionary sense; in other words, do all the bacterial cytochromes c arise from one common ancestral precursor? These evolutionary aspects of procaryotic cytochrome c will now be discussed on the basis of data concerning their primary or tertiary structure. Fig. 14 is a schematic representation of the primary structures from several bacterial cytochromes c. The sequences are aligned according to the haem-binding cysteines near the amino-terminal end of the molecule.

(A) Cytochrome c_2. Cytochrome c_2 has been purified from several *Athiorhodaceae* (non-sulphur photosynthetic purple bacteria), where it is probably a component of the photosynthetic electron-transfer chain. It resembles eucaryotic cytochrome c in its prosthetic haem group and redox potential ($+0.32$ V). However, the two cytochromes differ in their absorption spectrum and in their reactivity with mammalian cytochrome oxidase. Cytochrome c from *R. rubrum* and *R. molischianum* does not react at an appreciable rate with bovine and *Pseudomonas* cytochrome oxidase[227]. This may indicate that present-day cytochromes c_2, mitochondrial cytochrome c and *Pseudomonas* cytochrome c have undergone considerable evolutionary changes since they diverged from their common ancestors.

The primary[75] and tertiary[183] structures of cytochrome c_2 of *Rhodospirillum rubrum* have been determined. Although cytochrome c_2 from *Rhodospirillum rubrum* hardly reacts with bovine cytochrome oxidase, sequence homology between mammalian cytochrome c and cytochrome c_2 does exist when certain residues are deleted and others inserted. Horse heart cytochrome c and cytochrome c_2 from *R. rubrum* share 41 out of 112 amino acid residues (Fig. 15).

Sequence homology is particularly meaningful in areas of the polypeptide chain found to be invariant in a wide range of the taxonomic scale. Among the invariant residues are histidine and methionine (residues 18 and 91, c_2 numbering), both being engaged in coordinative bonds with the haem iron. The position of several acidic and basic amino acids as well as 10 of the 16 aromatic residues of cytochrome c_2 remained invariant in comparison with mitochondrial cytochrome c.

X-Ray diffraction analysis convincingly demonstrates that the tertiary

Fig. 15 — Alignment of cytochrome c (horse) and cytochrome c_2 (Rhodospirillum rubrum)

Block 1

Horse cytochrome c		- G D V E K G K K I F V Q K - C A Q C H T V E K G G K H K T G P N L H G L F G R K T
Rhodospirillum rubrum c_2		E G D A A A G E K - - V S K K C L A C H T F D Q G G A N K V G P N L F G V F E N T A
Common		G D G D V K C C H T G G K G P N L G F

(Upper numbering 1–40; lower numbering 1–40)

Block 2

Horse cytochrome c		G Q A P G F T Y T D A N - - - K N K G I T W K E E T L M E Y L E N P K K Y I P G
Rhodospirillum rubrum c_2		A H K D N Y A Y S E S Y T E M K A K G L T W T E A N L A A Y V K N P K A F V L E
Common		Y Y E T W K G L Y N P K

Block 3

Horse cytochrome c		- - - - - - T K M I F A G I K K K T E R E D L I A Y L K K A T N E
Rhodospirillum rubrum c_2		K S G D P K A K S K M T F - K L T K D D E I E N V I A Y L K - - T L K
Common		K M F K E I A Y L K T

Amino acid code

A alanine	F phenylalanine	L leucine
B asparagine or aspartic acid	G glycine	M methionine
C cysteine	H histidine	N asparagine
D aspartic acid	I isoleucine	P proline
E glutamic acid	K lysine	Q glutamine

R arginine	Y tyrosine
S serine	Z glutamine or glutamic acid
T threonine	- gap
V valine	
W tryptophan	

Fig. 15. Alignment of cytochrome c from horse and cytochrome c_2 from *Rhodospirillum rubrum*, according to Dus et al.[75]. Upper and lower numberings refer to horse cytochrome c and *R. rubrum* cytochrome c_2, respectively.

structure of bacterial cytochrome c_2[183] displays the same characteristic "cytochrome fold" as does mitochondrial cytochrome c[68,70], with minor residue deletions and two major chain insertions. Comparison of the tertiary structures[209] of cytochromes c and c_2 reveals that residues 56 and 76 (c numbering) are replaced by residues 56–59 and 79–87 (c_2 numbering), respectively. The latter insert interrupts the region of residues 70–80 (c numbering) which is invariant in all mitochondrial cytochromes c investigated. The nine inserted amino acid residues form a polar group projecting out of the protein and perhaps preventing the interaction with mammalian cytochrome oxidase. The position and the length of the insertions, inferred from the primary structure of cytochrome c_2 (see Fig. 15) agree rather nicely with the data from X-ray diffraction studies.

Comparison of the tertiary structures from cytochromes c and c_2 demonstrates that the absence of sequence homology at a•given site does not necessarily imply a lack of structural homology. Indeed, an aromatic residue always occurs in position 10 of the eucaryotic cytochrome c, whereas no aromatic amino acid is found at or near position 10 of bacterial cytochrome c_2. The tertiary structure, however, indicates that phenylalanine in position 20 of cytochrome c_2 has taken over the role of phenylalanine-10 (c numbering). Such a major change in primary structure probably could occur only because the two phenylalanine residues occupy almost identical spatial positions. This example shows the importance of both tertiary and primary structure analysis for significant conclusions regarding the evolutionary history of more distantly related proteins.

(B) Micrococcus cytochrome c-550. Micrococcus denitrificans can use molecular oxygen or inorganic nitrate as terminal electron acceptors. The primary structure of the respiratory cytochrome c-550 is not yet known but its tertiary structure has recently been determined to a resolution[209] of 4Å. The most striking feature, apparent from a comparison of cytochrome c-550 with eucaryotic cytochrome c and bacterial cytochrome c_2, is that all three redox proteins possess the same "cytochrome fold" although cytochrome c-550 is 33 residues longer than cytochrome c. α-Helical structures at the N- and C-terminus occur in the three cytochromes. The respiratory cytochromes (c-550 and c) resemble each other in the arrangement of aromatic residues in the "right" channel; they differ in this respect from the photosynthetic cytochrome c_2 of *Rhodospirillum*. However, in terms of gross chain folding cytochrome c-550 more closely resembles cytochrome c_2. Cytochrome c-550 indeed possesses both the inserted regions

(around position 50 and 70) found in cytochrome c_2, plus four inserted regions of a total of 25 amino acids[209]. These are all distributed on the surface of the molecule: five are added to the N-and C-terminus, five at the beginning of the carboxy-terminal helix and another ten in the region around position 20.

The cytochromes c, c_2 and c-550 with chain lengths of 104–137 amino acid residues are structurally very similar and presumably developed from a common cytochrome ancestor. These structural data suggest that the cyto-chrome c components of the photosynthetic (c_2-type) and respiratory (c- and c-550-type) redox mechanisms share an evolutionary ancestry. Timkovich and Dickerson[209] proposed a somewhat speculative evolutionary tree with an intermediate-size ancestral c_2-type cytochrome, leading to eucaryotic cytochrome c by deletion of 8 amino acid residues, and to c-550 cytochrome of *Micrococcus* by insertion of 25 amino acid residues.

(C) Cytochrome c-551 from pseudomonads. Pseudomonas cytochrome c-551 is an acidic, monohaem protein with 82 residues and a redox potential of $+0.28$ V. The primary structures of cytochrome c-551 from the following *Pseudomonas* species are known[3, 11]: *P. aeruginosa* P6009 (formerly called *P. fluorescens*), *P. stutzeri*, *P. mendocina* and *P. fluorescens* biotype C. In all four sequences more than 50% of the residues are identical, indicating that these cytochromes c are homologous with one another. The proline-rich region around the sixth iron-ligand methio-nine is maintained. Although these bacteria are all members of the same bac-terial genus, the differences between sequences from *P. aeruginosa* and *P. fluorescens* or *P. stutzeri* are as great as those between mitochondrial cytochrome c from mammals and insects, respectively (about 33%). The sequences of *Pseudomonas* cytochrome c-551 and mitochondrial cyto-chrome c differ considerably (Fig. 14). Different matching schemes have been proposed by several authors. Needleman and Blair[159] detected homologies between *Pseudomonas* cytochrome c-551 and a theoretically constructed eucaryotic ancestral cytochrome c[162]. In both proteins, 27 amino acids are identical and 11 are related by structural conservatism. Dickerson[69] proposed an interesting, theoretically constructed, three-dimensional model of a "pseudo-*Pseudomonas*" cytochrome c. The model was derived from the known tertiary structure of horse cytochrome c by insertions and deletions, including a 16-residue hairpin-loop in horse cytochrome c. However, definite proof or disproof of the homologous

nature of mitochondrial and *Pseudomonas* cytochrome c must await the unravelling of the tertiary structure of pseudomonad cytochrome c.

(D) Cytochrome c-553 from Desulfovibrio vulgaris. *Desulfovibrio* species contain at least three c-type cytochromes: c-553, c_3 and cc_3[35,36,131]. Cytochrome c-553 is a basic protein with a single haem group, 82 amino acid residues and a redox potential of -0.1 to 0 V. The primary sequence of cytochrome c-553[34] shows no homology with cytochrome c_3 from the same organism, but some similarities exist between it and the sequence of *P. fluorescens* cytochrome c-551. In both proteins the haem is attached near the N-terminus and sequence similarities exist in this region (Fig. 14). Both cytochromes contain a single histidine residue. The iron atom of the haem group is presumably coordinate-bonded to methionine 64 (c-553 numbering) or 61 (c-551 numbering).

(E) Cytochromes c$_4$ and c$_5$ from Pseudomonas. Large molecular weight c_4-type cytochromes were isolated from *Azotobacter vinelandii*[200,210], *Pseudomonas stutzeri*[126], *P. mendocina*[9] and *P. aeruginosa*[9]. They are acidic proteins with an α-band maximum at 552 nm, a redox potential of $+280$ mV and 2 haem groups per molecule. Cytochrome c_4 is a dihaem protein with a molecular weight of approximately 24000. Ambler and Murray[9] detected very similar N-terminal sequences (residues 1–20) for all the above-mentioned c_4 cytochromes. As in most other c-type cytochromes, one haem group is attached near the N-terminus. The sequence of the N-terminal heptapeptide of cytochrome c_4 shows some similarities to the N-terminus of *R. rubrum* cytochrome c_2 and *Chlorobium thiosulfatophilum* cytochrome c-555[9]. The very close similarity[9] between cytochrome c_4 from *A. vinelandii* and pseudomonads is surprising in view of the current classification of *Azotobacter*.

Cytochrome c_5 from *Pseudomonas mendocina* displays an α-band maximum at 555 nm, contains one haem per molecule and lacks tyrosine and phenylalanine. The primary structure of cytochrome c_5 from *P. mendocina*[10] is dissimilar from *Pseudomonas* cytochrome c-551. The segments immediately after the haem attachment sites of cytochrome c_5 and *Chlorobium* cytochrome c-555 are remarkably similar (Fig. 14). Other c_5-type cytochromes have been isolated from *Azotobacter vinelandii*[200,210] and several pseudomonads[10].

(F) Cytochrome c-555 from green sulphur photosynthetic bacteria. Cytochrome c-555 from *Chlorobium thiosulfatophilum* resembles algal cytochrome f in that it has 1 haem/molecule and a characteristically high $A\gamma/A\alpha$

absorption ratio of 7.1. Its absorption spectrum is similar to cytochrome
c-555 from "*Chloropseudomonas ethylica*", cytochrome c-556 from *Rhodopseudomonas palustris* (quoted by Meyer et al.[152]) and cytochrome c-554
from the blue-green alga *Anacystis nidulans*[230]. *Chlorobium* cytochrome c-
555 differs, however, from algal forms of c-type cytochrome f by its basic isoelectric point and comparatively low redox potential ($+0.14$ V). The reactivity of cytochrome c-555 and cytochrome f of several algae with *Pseudomonas* cytochrome oxidase was high[227], whereas cytochrome c-554 from *Anacystis nidulans* reacted poorly with the bacterial cytochrome oxidase[230]. The
primary structures of cytochromes c-555 from "*Chloropseudomonas ethylica*"
and *Chlorobium thiosulfatophilum*[217] are very similar. More than 50%
of the amino acid residues are invariant. Gray et al.[91] demonstrated that
cultures of "*Cps. ethylica*" are in fact associations of a green photosynthetic
bacterium, *Chlorobium limicola* and at least one sulphate-reducing heterotroph, containing cytochrome c_7 (see below, Section G). These results
strongly indicate that cytochrome c-555 from "*Cps. ethylica*" is in reality
cytochrome c-555 from *Chlorobium limicola*, present in the mixed culture of
"*Cps. ethylica*"[217].

  (G) Cytochrome c_3 and c_7. Cytochrome c_3 has been found in most
species of *Desulfovibrio* (anaerobic sulphate-reducing bacteria) and in some
green sulphur and non-sulphur purple photosynthetic bacteria[113]. Cytochrome c_3 plays a key role in the terminal reductive steps of sulphate
metabolism of *Desulfovibrio*. It differs in two major respects from all other
c-type cytochromes: it has a very low redox potential and contains four
haems per molecule. The amino acid sequences of cytochrome c_3 from
D. vulgaris[7] (61% GC), *D. gigas*[7] (60% GC) and *D. desulfuricans*[8] (55% GC)
have been reported. Apart from their haem-binding site, none of these
cytochromes c_3 is homologous to the other c-type cytochromes (except for
cytochrome c_7, see below). In cytochrome c_3, the first haem-binding site
is much further removed from the amino-terminal end of the molecule than
it is in all other c-type cytochromes. Cytochromes c_3 from *D. vulgaris* and *D.
gigas* have more than 50% of the amino acid residues in identical positions
and are obviously homologous proteins. A great number of important
deletions and insertions has to be made in order to align the four
haem-binding sites of *D. desulfuricans* cytochrome c_3 with cytochrome c_3
from *D. vulgaris* and *D. gigas* (Fig. 14). The third and fourth haem-binding
sites are closer together in *D. desulfuricans* cytochrome c_3 than in the
other two. The structure of the c_3-type cytochrome is peculiar because eight

cysteines are present; four occur as pairs in segments –Cys–A–B–Cys–His–
while the remaining four appear as pairs in segments which read
–Cys–A–B–C–D–Cys–His–(Fig. 14), except for the fourth haem-binding site
of *D. desulfuricans* which has the standard two residue segment for the
cysteines. These repeating structures may reflect internal homology in
cytrochrome c_3. The important structural differences in cytochrome c_3
between *D. desulfuricans* on the one hand and the two other *Desulfovibrio*
species on the other hand reflect the difference in % GC mentioned earlier.

Cytochrome c_7 $(c$-551.5$)^4$, isolated from "*Chloropseudomonas ethylica*",
is a new and structurally distinct type of cytochrome c possessing three
haems per molecule and three cysteine–histidine clusters similar to those of
mitochondrial cytochrome c. The marked similarities between cytochrome
c_7 and cytochrome c_3 from *D. desulfuricans* (Fig. 14) indicate that "*Cps.
ethylica*" cytochrome c-551.5 is a protein from the colourless sulphate-
reducing heterotroph present in the mixed culture "*Cps. ethylica*" (see section
F). This cytochrome c-551.5 is indeed absent from the photosynthetic
Chlorobium limicola, isolated from "*Cps. ethylica*"[91].

(H) Cytochrome c'. Cytochrome c'[117] constitutes still another class of
cytochromes c, characterized by the C-terminal position of the haem
group and a high-spin ligand field of the haem iron. Cytochrome c' has
been isolated from several photosynthetic bacteria. A haem-binding peptide
of cytochrome c' from *Chromatium* has been sequenced[73,117] (previously
called RHP or cytochrome cc'). Speculations and conclusions concerning
the evolution of c' and cc' cytochromes will be possible only when more
data concerning their primary structures have become available.

(3) Comparative biochemical studies on cytochrome c and its evolution.
Another important approach to the study of evolutionary relationships
between eucaryotic and procaryotic cytochromes c is based on comparison
of the relative reaction rates of various cytochromes c with nitrite reductase
from *Pseudomonas* and cow cytochrome oxidase[227–229]. In general, cytochromes c isolated from heterotrophic denitrifying bacteria react very
rapidly with *Pseudomonas* nitrite reductase, whereas those from higher
organisms react rapidly with cow cytochrome oxidase but poorly with the
Pseudomonas enzyme. These investigations may provide valuable information on the evolution of the functional properties of cytochrome c;
however, conclusions and speculations on the phylogeny of organisms, based
on the reactivities of their cytochromes c, are not always in agreement

with current thinking on the subject of evolution. This is not surprising because biological functions are very indirect translations of the genetic information incorporated in the nucleotide sequences of DNA.

We have summarized here the great range of complicated phenomena exhibited by the numerous combinations of haem c and protein in bacteria. That such a bewildering variety of cytochrome structures evolved in the bacterial world is not so surprising when one considers the widely different functions of these proteins in photosynthesis, respiration, sulphate reduction, etc. The differences in primary structure between, say, cytochromes c-551, c_3 and c' are so enormous that a polyphyletic origin would seem the simplest explanation[11]. It is indeed difficult to see how cytochromes with four haem groups at unusual positions (cytochrome c_3 and c') could possess the same "cytochrome fold" and hence the same evolutionary ancestor as, for example, monohaem cytochrome c_2 from *Rhodospirillum*.

It is to be expected that analysis of the tertiary and primary structures of many more bacterial cytochromes will yield significant conclusions on the evolution of the structure and function of these proteins in particular, and on bacterial groups in general. The main conclusion for the present is that many bacterial genera are at least as far apart in evolutionary terms as are animal phyla.

(ii) Iron–sulphur proteins

Iron–sulphur proteins[166] such as ferredoxins, rubredoxins and high-potential iron proteins, contain non-haem iron in a coordinate bond with cysteine sulphurs. Other common properties are: (*1*) the presence (except in rubredoxins) of a specific number of inorganic sulphur atoms, readily removable from the active center; (*2*) the ability to act as electron acceptor at specific redox potentials between -490 and $+350$ mV; (*3*) paramagnetism (characteristic EPR spectra) in one or more oxidation states and (*4*) acidity of the protein.

The iron–sulphur proteins are of considerable interest for unravelling and understanding biochemical evolution because these relatively simple electron carriers form a ubiquitous group of proteins occurring in almost all forms of life.

(1) Ferredoxin.
Ferredoxins are members of a group of iron–sulphur proteins occurring in a wide range of organisms such as the non-photo-synthetic anaerobic bacteria, the photosynthetic bacteria, the blue-green

TABLE IV

Characteristics of ferredoxins and ferredoxin-like proteins

Type of Fe–S cluster	Protein	Source	Approximate molecular weight	Redox potential (mV)	Type of reaction catalyzed	References
4 Fe + 4 S	8 Fe ferredoxin (2 centres)	Anaerobic bacteria (e.g. Clostridium)	6000	−390	phosphoroclastic reaction	22, 41, 67, 94, 134, 173, 207, 213
		Green photosynthetic bacteria	6000	ND	photosynthetic CO_2 fixation	40
		Chromatium	9000	−490	photosynthetic CO_2 fixation	17,184
	4 Fe ferredoxin (1 centre)	Desulfovibrio gigas	6000	ND	sulphate reduction	128, 129, 211
		Bacillus polymyxa	9000	−390	N_2 fixation	190, 234
	High-potential iron-sulphur protein (HiPIP) (1 centre)	Chromatium	9600	+350	Photosynthetic electron transport	76
2 Fe + 2 S	Plant ferredoxins	Blue-green algae, green algae and higher plants	10 500	−430	NADP photoreduction, N_2 fixation in blue-green algae	24, 52, 174, 231
	Adrenodoxin	Mammalian mitochondria	12 500	−270	hydroxylation of steroids	206
	Putidaredoxin	Pseudomonas putida	12 500	−240	hydroxylation of camphor	212

ND: not determined.

	Reference	1	2	3	4	5	6	7	8	9	1 0	1	2	3	4	5	6	7	8	9
(1) Clostridium pasteurianum	207	A	Y	K	I	A	D	S	C	V	S	C	G	A	C	A	S	E	C	P
(2) C. butyricum	22	A	F	V	I	N	D	S	C	V	S	C	G	A	C	A	G	E	C	P
(3) C. acidi-urici	173	A	Y	V	I	N	E	A	C	I	S	C	G	A	C	D	P	E	C	P
(4) C. tartarivorum	205	A	H	I	I	T	D	E	C	I	S	C	G	A	C	A	A	E	C	P
(5) Micrococcus aerogenes	213	A	Y	V	I	N	D	S	C	I	A	C	G	A	C	K	P	E	C	P
(6) Chromatium	144	A	L	M	I	T	D	Q	C	I	N	C	N	V	C	Q	P	E	C	P
(7) Desulfovibrio gigas	211	—	P	I	Q	V	D	N	C	M	A	C	Q	A	C	I	N	E	C	P
Common									C			C			C			E	C	P

	Reference	4 0	1	2	3	4	5	6	7	8	9	5 0	1	2	3	4	5	6	7	8	9
(1) Clostridium pasteurianum	207	C	—	G	N	—	—	—	—	—	—	—	—	C	A	N	V	C	P	V	G
(2) C. butyricum	22	C	—	G	N	—	—	—	—	—	—	—	—	C	A	N	V	C	P	V	G
(3) C. acidi-urici	173	C	—	G	A	—	—	—	—	—	—	—	—	C	A	G	V	C	P	V	D
(4) C. tartarivorum	205	C	—	G	A	—	—	—	—	—	—	—	—	C	Q	A	V	C	P	T	G
(5) Micrococcus aerogenes	213	C	—	G	S	—	—	—	—	—	—	—	—	C	A	S	V	C	P	V	G
(6) Chromatium	144	C	V	G	H	Y	E	T	S	Q	C	V	D	C	V	E	V	C	P	I	K
(7) Desulfovibrio gigas	211	L	—	D	D	—	—	—	—	—	—	—	—	Q	C	V	E	A	I	Q	S

Fig. 16. Alignment of bacterial ferredoxins (adopted from Orme-Johnson[166]). Amino acid code: see Fig. 15.

and green algae and the higher plants (for reviews see refs.[37,93,94,140,166,235]). They are the most electronegative electron carriers known, with redox potentials close to E'_0 of molecular hydrogen (-0.42 V at pH 7.55 for ferredoxin from C. pasteurianum[203]). Ferredoxins participate in

2 0	1	2	3	4	5	6	7	8	9	3 0	1	2	3	4	5	6	7	8	9
V	N	A	I	S	Q	G	D	S	I	F	V	I	D	A	D	T	C	I	D
V	S	A	I	T	Q	G	D	T	Q	F	V	I	D	A	D	T	C	I	D
V	D	A	I	S	Q	G	D	S	R	Y	V	I	D	A	D	T	C	I	D
V	E	A	I	H	E	G	T	G	K	Y	Q	V	D	A	D	T	C	I	D
V	N	—	I	Q	Q	G	—	S	I	Y	A	I	D	A	D	S	C	I	D
N	G	A	I	S	Q	G	D	E	T	Y	V	I	E	P	S	L	C	T	E
V	D	V	F	Q	M	D	E	Q	G	D	K	A	V	N	I	P	N	S	N

6 0	1	2	3	4	5	6	7	8	9	7 0	1	2	3	4	5	6	7	8	9	8 0	1
A	P	V	Q	E	—																
A	P	N	Q	E	—																
A	P	V	Q	A	—																
A	V	K	A	E	—																
A	P	N	P	E	D																
D	P	S	H	E	E	T	E	D	E	L	R	A	K	Y	E	R	I	T	G	E	G
C	P	A	A	I	R	S															

Identical positional residues in ferredoxins (1)–(6) are underlined and cysteine molecules are boxed

a wide variety of low-potential electron-transfer processes such as the production and utilization of hydrogen gas by anaerobic bacteria (phosphoroclastic pyruvate cleavage), nitrogen fixation and photosynthesis; in these processes they act as electron carrier between the photoreduction

References p. 72

system and NADP. Reduced ferredoxin from photosynthetic bacteria is involved in CO_2 fixation and in reactions of the reductive carboxylic acid cycle[38,39,80]. Ferredoxin also functions as an electron carrier in the reduction of urate to xanthine by purinolytic clostridia[215].

Depending on their iron–sulphur content ferredoxins may be classified into two main types: (*1*) the 4Fe + 4S bacterial-type ferredoxin, containing one or two tetrameric ion–sulphur clusters and (*2*) the 2Fe + 2S plant-type ferredoxins. Some important characteristics of these proteins are summarized in Table IV.

These ferredoxins share several properties: (*a*) a low redox potential (from -0.39 V to -0.49 V at pH 7.0); (*b*) they contain equivalent amounts of iron and inorganic acid-labile sulphur, (*c*) they are strongly acidic, (*d*) they are functionally interchangeable in certain biological reactions such as the photoreduction of NADP by chloroplasts. They differ, however, in spectral characteristics, in the number of iron–sulphur groups, in amino acid composition and in molecular weight.

The complete amino acid sequences of five chloroplast-type ferredoxins are known (*Scenedesmus*[52], spinach[52], alfalfa[52], *Colocasia esculenta*[174], *Leucaena glauca*[24]) and of seven microbial ferredoxins (*C. pasteurianum*[207], *C. butyricum*[22], *C. acidi-urici*[173], the thermophile *C. tartarivorum*[205], *Micrococcus aerogenes*[213], *Chromatium*[144] and *Desulfovibrio gigas*[211]. The sequences of bacterial ferredoxins and the sequence of the high potential iron protein (HiPIP) from *Chromatium*[76] are aligned in Fig. 16. Several gaps had to be introduced for the non-photosynthetic bacteria in order to align them with the purple sulphur bacterium *Chromatium* (81 residues).

(A) The 8 Fe + 8 S ferredoxins. Thus far 8 Fe + 8 S ferredoxins have been only isolated from photosynthetic and non-photosynthetic bacteria, *e.g.* the photosynthetic green sulphur *Chlorobium* and the purple sulphur *Chromatium*, the anaerobic clostridia and peptococci. These bacterial-type ferredoxins show a characteristic absorption peak at 390 nm. They are relatively small proteins of about 55 amino acids and they have a molecular weight of approximately 6000. The invariance of the 8 cysteine positions (Fig. 16) has been correlated with the results of crystallographic studies, demonstrating two identical tetrameric iron–sulphur clusters where each Fe atom forms a coordinate bond with three inorganic sulphurs and one cysteine residue[109]. The bacterial 8 Fe ferredoxins are two-electron carriers with redox potentials close to the E'_0 of the hydrogen electrode (-420 mV). This property makes these proteins suitable for operating under a primitive,

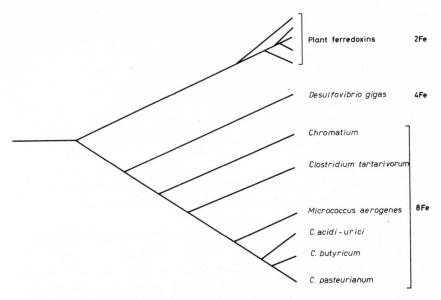

Fig. 17. Possible phylogenetic relationships between plant and bacterial ferredoxins.

reducing atmosphere.

The ferredoxins of *Micrococcus* and the four clostridia have 26 out of 54–55 amino acids in identical positions (Fig. 16), indicating that approximately 50% of the protein remained unchanged in the course of evolution. Sixteen amino acid positions are invariant when *Chromatium* ferredoxin is compared with the compound from the clostridia and *Micrococcus*. Only three amino acids out of 55 are in identical positions when bacterial and plant-type ferredoxins (96–97 amino acid residues) are compared.

Dayhoff[52] derived an evolutionary tree of ferredoxins from reconstructed ancestral sequences. With the new data for *C. tartarivorum*[205] and *D. gigas*[211], a possible evolutionary tree is obtained as represented in Fig. 17. The similarity between ferredoxins from the saccharolytic clostridia *C. pasteurianum* and *C. butyricum* respectively is greater than that between ferredoxins from a saccharolytic and a purinolytic *Clostridium* (*C. acidiurici*). This is in agreement with the known metabolic and taxonomic differences between these organisms[56].

The structure of *Chromatium* ferredoxin is unique in that it possesses a large acidic amino acid cluster near the carboxyl-terminal region. The

References p. 72

```
                                 1                   2
             1 2 3 4   5 6 7 8 9 0 1 2 3 4 5 6 7 8 9 0 1 2 3 4 5 6 7 8

1st half ( 1–28)   A Y K I — A D S |C| V S |C| G A |C| A S E |C| P V N A I S Q G D S
2nd half (29–55)   I F V I D A D T |C| I D |C| G N |C| A N V |C| P V G A P V Q E — —

                 3                   4                   5
             9 0 1 2 3 4 5 6 7 8 9 0 1 2 3 4 5 6 7 8 9 0 1 2 3 4 5

Common             I   A D   C       C G   C A     C P V   A     Q
```

Fig. 18. Alignment of the half-chains of ferredoxin from *C. pasteurianum* (from Dayhoff[51a]). Amino acid code: see Fig. 15. Upper and lower numbering refer to first and second half-chains of ferredoxin, respectively.

possibility must not be ruled out that this unique structure fulfills specific functions (*e.g.* in photosynthetic mechanisms) which are lacking in other bacterial ferredoxins[144].

It is possible to construct an interesting hypothesis on the early evolution of ferredoxin by comparing the amino acid sequences in one single type (*e.g. C. pasteurianum*). When the sequence 29–55 is aligned with the segment 1–28 and a gap is introduced between residues 4 and 5 (Fig. 18), it is seen that 13 (*i.e.* 46%) of the residues in the two half-chains are identical[52,77]. This suggests that the half-chains are homologous and that the protein arose by gene duplication. The ferredoxins from the clostridia and *M. aerogenes* contain four cysteine residues in each half of the molecule, separated by regular interspaces of 2, 2 and 3 amino acids, followed by a proline-valine sequence. Eck and Dayhoff[77] proposed that each half of the clostridial ferredoxin molecule was built up from a smaller peptide [Ala–Asp(or Pro)–Ser–Gly] by a doubling process, occasionally interrupted by mutation. Comparison of the primary structures of the four clostridial ferredoxins (Fig. 16) reveals[93] that only 9 of the 20 amino acids are present in all four proteins. *C. butyricum* ferredoxin may be considered to be the simplest ferredoxin sequence so far because these 9 amino acids (glycine, alanine, valine, glutamic acid, aspartic acid, proline, cysteine, isoleucine and serine) comprise 91% of the molecule and have all been synthesized in the laboratory under simulated primitive-earth conditions[93]. Moreover, six of these nine amino acids (glycine, alanine, valine, glutamic acid, aspartic acid and proline) were detected in the Murchison meteorite[127] and constitute 64% of the total amino acid content of *C. butyricum* ferredoxin. It should also be noted that these ferredoxins contain no or very little lysine, histidine, arginine, tryptophan and methionine. Hall *et al.*[93] suggested that under primitive-earth conditions a ferredoxin prototype may have been formed as a short polypeptide chain of 26 amino acids, containing the above-mentioned 9 amino acids. Inorganic iron and sulphur existed in abundance and could easily have been incorporated into the primitive apoprotein by a non-enzymatic process to form an active ferredoxin molecule. The latter step was indeed performed in the laboratory with fragments of the apoprotein of *C. acidi-urici*[167].

(B) The 4 Fe + 4 S ferredoxins. Ferredoxins containing 4 Fe + 4 S per mole have recently been discovered in *(1) Desulfovibrio gigas*[211] an obligate anaerobe, where ferredoxin plays a role in the sulphate reduction; *(2) Bacillus polymyxa*[190, 234], a facultative nitrogen fixer, where ferredoxin transfers electrons from pyruvate to the nitrogenase system, and *(3) Spirochaeta aurantia*[112], a facultative anaerobe.

The 4 Fe ferredoxins are one-electron carriers with a molecular weight of 6000–9000 (Table IV). The primary structure of *D. gigas* ferredoxin[211] in its first half (residues 1–29) shows a high degree of homology with the clostridial (8 Fe) ferredoxins, whereas the second half contains only two cysteine residues and shows in some segments a high degree of homology with the 2 Fe ferredoxins of green plants. Consequently, an attractive hypothesis[94] is that the 4 Fe ferredoxins are evolutionary links between the 8 Fe and the 2 Fe ferredoxins (see Fig. 19).

(C) The 2Fe + 2S ferredoxins. The "plant-type" ferredoxins contain two iron and two labile sulphur atoms forming coordinate bonds to four cysteine residues. Ferredoxins from both blue-green algae (*Anacystis nidulans, Anabaena variabilis* and *Nostoc muscorum*) and chloroplasts have spectra with absorption peaks of the oxidized form at 276, 330, 420 and 463 nm. The 2Fe ferredoxins are more widespread than was previously thought: not only are they present in blue-green algae, green algae and higher plants, but also as recently demonstrated, in aerobic and anaerobic bacteria (*E. coli*[219], *Pseudomonas putida*[212], *Rhodopseudomonas rubrum*[187] and *Azotobacter vinelandii*[65]) and in beef and pig adrenals (adrenodoxin)[206].

The 2Fe ferredoxins from plants are one-electron transfer proteins. Almost twice as long as those from the nonphotosynthetic bacteria, they have a molecular weight of about 10 500 and a redox potential of about −430 mV (Table IV). The evolutionary tree derived from the amino acid sequences of ferredoxins from five plants (one green alga, one mono-cotyledon and three dicotyledons) corresponds well with the classification of these plants[174]. Both plant and bacterial ferredoxin display a certain degree of homology, suggesting a common ancestor.

The 2Fe proteins adrenodoxin from adrenals and putidaredoxin from
P. putida are components of the electron-transport systems which
hydroxylate steroids and camphor respectively. Both proteins were
sequenced[206, 212], contain 114 amino acids and show clear-cut homologous
sequences at the amino-terminus and around the two cysteine residues[212],
but cannot be substituted for one another in reconstitution experiments.
Both exhibit only a fair degree of sequence homology with the 2Fe plant and
8Fe bacterial ferredoxins, indicating that an evolutionary relationship, if any,
with the latter proteins can only be a very distant one.

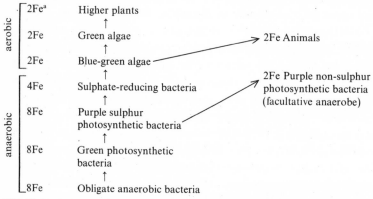

Fig. 19. Evolutionary development of ferredoxins (adopted from Hall *et al.*[94]). [a] Number
of Fe atoms (and the equivalent number of S atoms) per mole of ferredoxin.

On the basis of a comparative study of the primary structure and
properties of ferredoxins from a great variety of organisms Hall *et al.*[94]
proposed a possible evolutionary development of the ferredoxins as represented in Fig. 19. In this scheme the two-electron transfer ferredoxins
with two 4Fe + 4S clusters are regarded as the earliest ferredoxins that
evolved from an extent primitive iron–sulphur protein. This primitive
ferredoxin may have been formed as outlined in section *A* (p. 56). Later
on, one of these 4Fe + 4S clusters was lost and the one-electron-transfer
ferredoxins evolved, occurring at present in *D. gigas* and *B. polymyxa*.
The 4Fe + 4S ferredoxins may be considered to be the precursors of the
2Fe + 2S ferredoxins, with longer amino acid chains, occurring in algae and
higher plants. Parallel to the evolution of the plant-type ferredoxins,

proteins similar to the ferredoxins were incorporated into the metabolic pathways of animals where they participate in electron-transfer reactions.

Ferredoxins and cytochrome c_3 (see p. 48) both have low redox potentials and can substitute for each other functionally in some reactions[2]. The ferredoxins contain sequence segments of the type Cys–A–B–Cys, Cys–A–B–C–Cys and Cys–A–B–C–D–Cys (plant-type ferredoxin), which are similar to the haem-binding sites of cytochromes c_3. A challenging hypothesis is that ferredoxins and cytochromes are very distantly related proteins.

As more data on the primary and tertiary structures of other ferredoxins and ferredoxin-like proteins will become available many more interesting conclusions may be expected on the evolution pattern of this class of proteins.

(2) High-potential iron protein (HiPIP). An unusual iron–sulphur protein, containing four cysteine residues, four iron atoms and four labile sulphur atoms, has been isolated from three purple photosynthetic bacteria: *Chromatium*[74, 76], *Rhodopseudomonas*[54] and *Thiocapsa*[116]. The protein from *Chromatium* has a molecular weight of 9600 and a redox potential of +350 mV (Table IV); it plays a role in the photosynthetic electron transport chain[81]. Its primary structure[76] shows no obvious relationships to those of other iron–sulphur proteins.

HiPIP is non-magnetic in the reduced state whereas all known reduced ferredoxins are magnetic. In spite of the large difference in redox potential between HiPIP and ferredoxin (740 mV), X-ray analysis of HiPIP[44] revealed no structural difference between the 4Fe+4S cluster from HiPIP and the two 4Fe+4S clusters of the 8Fe ferredoxins. A plausible explanation for this apparent contradiction was given by Hall *et al.*[94]. By using mild denaturing agents such as dimethylsulphoxide, they were able to reduce the reduced HiPIP to a "superreduced" state which shows a redox potential like that of reduced ferredoxin (−390 mV). The 4Fe+4S cluster can thus assume redox potentials of either +350 mV or −390 mV, depending upon its association with different types of apoprotein. Organisms equipped with this particular 4Fe+4S cluster perhaps enjoyed great selective advantages in the course of their evolution.

(3) Unclassified ferredoxins. *Azotobacter vinelandii*, an aerobic nitrogen-fixing bacterium, contains two types of ferredoxins[232, 233]. Both proteins

Upper numbering (1–29): 1 2 3 4 5 6 7 8 9 | 10 1 2 3 4 | 5 6 7 8 9 | 20 1 2 3 4 5 6 7 8 9

	Reference	Sequence (boxed = common cysteine; identical positional residues underlined in original)
(1) Clostridium pasteurianum[a]	148	fM K K Y T [C] T V [C] G Y I Y D — P E D G D P D D G V D P G T D
(2) Peptostreptococcus elsdenii	14	M D K Y E [C] S I [C] G Y I Y D — E A E G D — D G N V A A G T K
(3) Micrococcus aerogenes	16	M Q K F E [C] T L [C] G Y I Y D — P A L V G P D T P D Q D G — A
(4) Pseudomonas oleovorans (N)[b]	23	— A S Y K [C] P D [C] N Y V Y D — E S A G N V H E G F S P G T P
(5) Pseudomonas oleovorans (C)[b]	23	— L K W I [C] I T [C] G H I Y D W E A L G D E A E G F T P G T R

Lower numbering (119–174): 129 130 131 132 133 134 135 136 137 138 139 | 140 141 142 143 144 145 146 147 148 149 | 150 151 152 153 154 155 156 157

	Common
Common	C ... C ... Y D ... G

Fig. 20. Amino acid sequences of rubredoxins (adopted from McCarthy[148]). Amino acid code: see Fig. 15. The upper numbering (1–54) refers to proteins (1)–(4); the lower numbering (119–174) to the COOH-terminal residues (5) of *P. oleovorans* rubredoxin. Identical positional residues are underlined and common cysteine molecules are boxed. [a] Rubredoxin of *C. pasteurianum* is *N*-formylated at the NH$_2$-terminal methionine. [b] *P. oleovorans* (N) and (C) are the NH$_2$- and COOH-terminal portions, respectively, omitting residues 55–118 to emphasize homologous regions.

are effective electron carriers for the nitrogenase reaction. *Azotobacter* ferredoxins FdI and FdII both contain eight iron and eight sulphur atoms and have a redox potential of -240 and -460 mV respectively. *Azotobacter* FdI has a molecular weight of 14 500, which is more than that of any ferredoxin previously described. Yoch and Arnon[232] suggested that the large size of this ferredoxin is perhaps an evolutionary adaptation enabling this strictly aerobic micro-organisms to conserve the strong reducing power needed for nitrogen fixation in an aerobic environment.

Similarly, *Rhodospirillum rubrum* appears to contain two types of ferredoxin[188]: type I has six iron and six sulphur atoms and a molecular weight of 8700, whereas type II has two iron and two sulphur atoms and a molecular weight of 7500. It thus appears that in some bacteria ferredoxins, like cytochromes, occur in several forms.

(4) Rubredoxin. Rubredoxins[166] are classified as Fe–S proteins because the Fe atom is chelated to four cysteine sulphurs as it is in the ferredoxins. Unlike the ferredoxins, however, these proteins contain no acid labile sulphur and only a single iron atom which functions as electron acceptor[15,97,136]. Rubredoxins have been purified from the anaerobic

```
3                        4                            5
0   1 2 3 4 5 6 7 8   9 0 1 2      3 4    5 6 7 8 9 0 1 2 3 4

F   K D I P D D W V  C  P L  C  — — G V — G K D E F E E V E E
F   A D L P A D W V  C  P T  C  — — G A — D K D A F — V K M D
F   E D V S E N W V  C  P L  C  — — G A — G K E D F E V Y E D
W   H L I P E D W D  C  P C  C  — — A V R D K L D F M L I E S
F   E N I P — D W D  C  C W  C  B P G A — T K E N Y V L Y E E K


        1                    1                        1
        5                    6                        7
8   9 0 1 2    3 4 5 6 7 8 9 0 1 2 3    4 5 6 7 8 9 0 1 2 3 4
        W          C         C                   K
```

bacteria *Clostridium pasteurianum*[135], *Micrococcus aerogenes*[13], the rumen bacterium *Peptostreptococcus elsdenii*[14], sulphate-reducing bacteria *Desulfovibrio gigas*[128,129], *D. desulfuricans*[160], *D. vulgaris*[33] and two species of green sulphur photosynthetic bacteria, *Chlorobium thiosulphatophilum* and "*Chloropseudomonas ethylica*"[153]. Rubredoxin was not, however, detected in the purple photosynthetic bacteria *Chromatium vinosum, Rhodospirillum rubrum* and *Rhodopseudomonas palustris*[153].

The absorption spectra of the oxidized rubredoxins from the anaerobic bacteria show peaks at 490, 380 and 280 nm. They are strongly acidic proteins, characteristically lacking arginine and histidine; they have a molecular weight of approximately 6000. While the specific role of rubredoxin is not yet understood it appears to function as a one-electron carrier in redox reactions[135]. The redox-potential ($E'_0 = -0.06$ V at pH 7.0[135]) is much higher than the corresponding value (-0.42 V) for bacterial ferredoxin. *In vitro* rubredoxin can replace ferredoxin as electron carrier in certain enzymatic reactions[135].

Rubredoxin, isolated from the strict aerobe *Pseudomonas oleovorans*, is involved in ω-hydroxylation reactions of fatty acids and hydrocarbons[169]. The protein exhibits absorption maxima at 280, 377 and 495 nm

References p. 72

and probably contains two iron atoms/molecule; its molecular weight of 19 000 is larger than that of all other rubredoxins.

The amino acid residues 1–53 and 119–173 of rubredoxin from *Pseudomonas oleovorans*[23] can be aligned[166] with the rubredoxins from *Clostridium pasteurianum*[148], *Peptostreptococcus elsdenii*[14] and *Micrococcus aerogenes*[16] (Fig. 20). The primary structure of rubredoxin from *Clostridium pasteurianum* was recently inferred from the electron density map at 2 Å resolution[97]. The alignment of the rubredoxins (Fig. 20) shows clear-cut similarities between all the sequences under comparison, *e.g.* the cysteine molecules occur at the same positions with respect to the amino terminal end. Similarities between rubredoxins from the three anaerobic bacteria (*C. pasteurianum, P. elsdenii* and *M. aerogenes*) are greater than those between the two residues 1–53 and 119–173 of the *P. oleovorans* rubredoxin (see difference matrix, Table V). No apparent homologies are detectable between the center part (residues 54–118) of the *P. oleovorans* rubredoxin and the other halves of the molecule.

TABLE V

Difference between rubredoxins from *C. pasteurianum, P. elsdenii, M. aerogenes,* and NH_2- and COOH-terminal parts of rubredoxin from *P. oleovorans,* in percentages.

	C. pasteurianum	*P. elsdenii*	*M. aerogenes*	*P. oleovorans* (N)	*P. oleovorans* (C)
Clostridium pasteurianum	0				
Peptostreptococcus elsdenii	46	0			
Micrococcus aerogenes	48	52	0		
Pseudomonas oleovorans (N)[a]	60	65	73	0	
Pseudomonas oleovorans (C)[a]	58	61	63	60	0

[a] *P. oleovorans* (N) and (C) are the NH_2- and COOH-terminal portions; residues 55–118 are omitted.

Benson *et al.*[23] suggested that the various rubredoxins arose from one common ancestor and that gene duplication took place during the evolution of the rubredoxin from the strict aerobe *Pseudomonas oleovorans*. The possible phyletic relationships between the various rubredoxin-containing bacteria are summarized in Fig. 21.

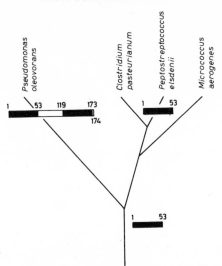

Fig. 21. Evolutionary tree of rubredoxin (Benson *et al.*[23], modified). Length of polypeptide chain and homologous regions (black) are represented schematically.

The rubredoxins from *C. pasteurianum* and *M. aerogenes* are less similar (48% difference, Table V) than their corresponding ferredoxins (30% difference[52]), indicating that during the evolution of these two bacteria the number of accepted point mutations[52] was greater on the rubredoxin gene than on the ferredoxin gene.

Comparison of the amino acid residues 13–36 of *Scenedesmus* or spinach ferredoxin and residues 1–23 of *M. aerogenes* rubredoxin reveals a more than fortuitous number of similarities, suggesting an archetypal correlation between bacterial rubredoxin and plant ferredoxin[221] (see Fig. 22). It is thus quite possible that ferredoxin and rubredoxin have a common ancestor. The rubredoxins and ferredoxins promise to become important guides for future phylogenetic work in the area of iron–sulphur proteins.

	1								2									3						
	3	4	5	6	7	8	9	0	1	2	3	4	5	6	7	8	9	0	1	2	3	4	5	6
Scenedesmus ferredoxin (13–36)	N	Q	T	I	E	C	P	D	D	T	Y	I	L	D	A	A	E	E	A	G	L	D	L	P
M. aerogenes rubredoxin (1–23)	M	Q	K	F	E	C	T	L	C	G	Y	I	Y	D	P	A	L	V	—	G	P	D	T	P

									1										2			
	1	2	3	4	5	6	7	8	9	0	1	2	3	4	5	6	7	8	9	0	1	2	3
Common		Q			E	C						Y	I		D		A				G		D	P

Fig. 22. Similarities between *Scenedesmus* ferredoxin and *Micrococcus aerogenes* rubredoxin[221]. Amino acid code: see Fig. 15. Upper and lower numbering refer to residues of *Scenedesmus* ferredoxin and *M. aerogenes* rubredoxin respectively.

References p. 72

TABLE VI

Distribution, chemical characteristics and biological activities of flavodoxins

Micro-organism	Molecular weight	Spectral characteristics of oxidized form (maxima in nm)			Redox potentials (mV)	Number of amino acid residues	Amino acids absent	Biological activities		References
								Nitrogen fixation	Phosphoro-clastic reaction	
FERMENTATIVE BACTERIA										
Clostridium pasteurianum	14600	272	372	443	−132, −419	148	his	+	+	124, 125, 137
Clostridium MP	13800	272	376	445	−90, −399	N.D.	N.D.	N.D.	+	145
Peptostreptococcus elsdenii	15000	272	377	445	−115, −373	138	his	N.D.	+	146, 147
Desulfovibrio gigas	16000	273	374	456	N.D.	149–150	his	N.D.	+	72, 130
Desulfovibrio vulgaris	16000	273	375	456	N.D.	151–152	met	N.D.	+	72
Escherichia coli K 12	14500	274	370	467	−285, −455	120–121	none	+	+	219
PHOTOSYNTHETIC BACTERIA										
Rhodospirillum rubrum	23000	272	376	460	N.D.	210	none	N.D.	N.D.	50
AEROBIC BACTERIA										
Azotobacter vinelandii	23000	N.D.			−270, −460	186	cys, his	+	−	78, 191, 218

N.D.: not determined

(iii) Flavodoxin

Flavodoxin was first isolated as an electron-transfer flavoprotein from iron-deficient cells of *Clostridium pasteurianum*[124,125]. It substitutes for the iron–sulphur ferredoxin in the anaerobic oxidation of pyruvate (phosphoro-clastic reaction) and other ferredoxin-dependent reactions catalyzed by extracts of both *C. pasteurianum* and *Peptostreptococcus elsdenii*. Studies of the oxidation/reduction properties of flavodoxin from *P. elsdenii* have shown that reduction of the flavin can occur in two one-electron steps with E'_0 values of -0.115 and -0.373 V at pH 7.0[146]. Table VI summarizes some features of the known flavodoxins. For recent reviews, see refs. 199, 235.

Flavodoxins contain one mole FMN per mole of protein. Like the ferre-doxins, they transfer electrons at low redox potentials and possess an acidic protein backbone. Unlike the ferredoxins, however, they are devoid of metals and do not contain labile sulphide. Flavodoxin is yellow in the oxidized form with absorption maxima at 272, 372 and 443 nm and a shoulder at 472 nm[124]. The addition of apoprotein to FMN modifies the spectrum. When compared with FMN, both *Desulfovibrio* and *E. coli* flavodoxins exhibit an important shift around 450 nm (10 and 20 nm respectively) towards the red. *C. pasteurianum* displays a 2-nm shift towards the blue end of the spectrum, and flavodoxin from *P. elsdenii* shows no shift at all.

The properties of the flavodoxins from the clostridia, *P. elsdenii*, two *Desulfovibrio* species and *E. coli* show considerable overall similarities, in-dicating an evolutionary relationship between these proteins. They have a molecular weight of 14000–16000 with 120–152 amino acid residues. A high percentage of these amino acids is acidic, as in ferredoxin and rubredoxin. Flavodoxins have 0–1 histidine residues with the exception of *E. coli*, possessing 3 histidine residues. Although cysteine appears to be involved in the binding of FMN, important differences exist in the number of cysteine residues: flavodoxins of sulphate-reducing bacteria contain five of them against one or two for the other organisms and none for the flavodoxin from *Azotobacter*. Flavodoxin of *D. vulgaris* differs from all others by the absence of methionine (Table VI). Comparison of residues 1–80 from the known primary structures of flavodoxin from *P. elsdenii*[204] and *C. pas-teurianum*[83] reveals a high degree of structural homology. There is no ob-vious similarity between the primary structures of flavodoxin and clostridial ferredoxin[83].

The molecular weights of the flavodoxins from *Rhodospirillum rubrum* and *Azotobacter vinelandii* are considerably larger (23000 dalton, Table VI) than those of the corresponding proteins from the fermentative bacteria (approximately 15000 dalton). If they are at all related to the flavodoxins from the latter bacteria, then only distantly. Flavodoxin from the obligate aerobic nitrogen-fixer *Azobacter* participates as an electron carrier in N_2 fixation together with *Azotobacter* ferredoxin (see p. 59). Flavodoxin from *Azotobacter* (also named azotoflavin or Shethna protein) is unique, because its semiquinone is not readily reoxidized[21]. Because the flavodoxins from the aerobic *Azotobacter* are larger than those from the clostridia, they may reflect an evolutionary trend toward low redox potential electron-transfer proteins, operating in an aerobic environment.

(iv) Thioredoxin

Thioredoxin from *E. coli* is a small acidic protein (molecular weight 12000, 108 amino acid residues) which participates as hydrogen donor in the enzymatic reduction of ribonucleotides to deoxyribonucleotides. Its primary structure[100] bears very little resemblance to either ferredoxin or rubredoxin. It is therefore possible that the analogous spacing of the cysteines (Cys–Gly–Pro–Cys) is convergent. Thioredoxin has been purified from *Saccharomyces cerevisiae*[171]. As expected, amino acid composition and peptide maps after tryptic digestion of thioredoxin from both organisms are considerably different. The total number of amino acid residues of yeast thioredoxin is 109–111, and its molecular weight is 12600.

(v) Azurin

The blue bacterial protein azurin is one of the simplest copper proteins known. Common features of the azurins are the low molecular weight (15000), the presence of one Cu atom per molecule and the high optical absorption around 600 nm. Their absorption spectrum is almost identical with other, more complex copper proteins such as ceruloplasmin (molecular weight 140000, 8 Cu atoms). *In vitro* azurin will accept electrons from *Pseudomonas* cytochrome *c*-551.

The amino acid sequences of azurins from the following Gram-negative aerobic bacterial species are known[52]: *Pseudomonas fluorescens*[6], *Bordetella bronchiseptica*[52], *Alcaligenes denitrificans*[52] and *A. faecalis*[52]. The primary structure of the azurins (128–129 amino acid residues) from representatives of three different bacterial genera are clearly homologous (for alignment of azurins: see Dayhoff[52]). The difference matrix[52] shows that the evolutionary distance between the four proteins is almost equal. This situation seems paradoxical because two of these azurins belong to species from the same genus *Alcaligenes*. This genus, however, is poorly defined taxonomically.

Kelly and Ambler[115] detected sequence similarity around the single thiol group of bacterial azurin and plastocyanine from the green alga *Chlorella fusca*. The similar concentration of aromatic residues in the vicinity of the cysteine residue reflects at least a functional similarity between both copper-containing proteins.

(vi) Nuclease

Nucleases of *Staphylococcus aureus* strains V8 and Foggi contain 149 amino acid residues and have been sequenced[49, 51, 208]. Both nucleases differ in a single amino acid residue: leucine in the 124th position of the V8 strains is replaced by histidine in the Foggi strain.

(b) Electrophoretic analysis of cellular proteins and enzymes

The measurement of phylogenetic relationships, obtained by comparison of the primary structures of homologous proteins requires delicate and laborious sequencing techniques on highly purified proteins.

Zone electrophoresis rapidly yields comparisons of large numbers of proteins and enzymes from different micro-organisms. One compares the number of protein bands, their relative mobilities (distance travelled from the origin) and relative concentrations. The electrophoretic behaviour of proteins on polyacrylamide gel is determined by their charge, shape and dimensions (molecular weight). One of the advantages of this method is that some minute structural differences between proteins, and hence between cistrons, can be detected easily without their nature being understood, however. It should be stressed here that identical electrophoretic mobilities for proteins from different organisms do not necessarily imply that these proteins are identical.

Some authors suggested that polyacrylamide-gel electropherograms of cell proteins from various organisms can be arranged to represent genetic relatedness. This method is very valuable for the clarification of some taxonomic problems concerning micro-organisms (see below), but it is highly dubious that it can reveal a phylogenetic tree. Indeed, of the 20 amino acid residues found in proteins eight are electrically neutral and interchange amongst them usually does not alter electrophoretic mobility. Substitution of, say, a lysine residue by arginine will have little effect on the mobility of the protein whereas substitution of an acidic residue by a neutral or basic one can cause a drastic change in electrophoretic behaviour. It is obvious that a mutational change in the genome will not always cause alterations in the electrophoretic properties of a given protein. It is our experience, based on combination of the results from electrophoretic methods on the one hand and DNA–DNA hybridizations on the other hand, that similar protein profiles permit the conclusion that two or more organisms are identical or very similar; differences in profile, however, give no clue as to the degree of phylogenetic remoteness.

Publications on the application of zone electrophoresis of cell proteins in bacterial taxonomy and microbial evolution have increased considerably during the last ten years. Some important contributions will be briefly discussed here. A review on this subject has appeared[84].

(i) Protein profiles

Sacks *et al.*[181] compared phenol–acetic acid extracts from 84 strains of *Enterobacteriaceae* by disc electrophoresis. All isolates showed similar electrophoretic patterns, all considerably different from the unrelated bacteria *Staphylococcus* and *Pseudomonas*. The dose similarity between the protein profiles of *E. coli* and 14 *Shigella* strains was in good agreement with their close genetic relatedness as evidenced by DNA–DNA hybridizations (Brenner *et al.*, see p. 8). There is good correlation between the protein profiles and the accepted biochemical classification of these bacteria. Nevertheless, the following discrepancies were apparent: lack of uniformity of pattern within the *Citrobacter* group, and similarity between the *Salmonella* and the *Serratia* strains.

Systematic analysis of electrophoretic protein profiles has proven to be a very dependable tool for the taxonomy and identification of the *Mycoplasmatales*. The morphology of these organisms is subject to great variations and their biochemical activity is weak. As a result, the classification of these micro-organisms is extremely difficult and based largely on their serological behaviour. Good correlation exists between the results of polyacrylamide-gel electrophoresis of cell proteins of mycoplasmata, their DNA base composition and serological reactions[175–177, 179, 214]. Several *Mycoplasma hominis* strains *e.g.* showed marked differences corresponding to their known serological and DNA dissimilarities. Recent descriptions of *Mycoplasma* species are partly based on their protein profiles on polyacrylamide gels. Tully *et al.*[214] *e.g.* demonstrated by serological, electrophoretical and biochemical techniques that *M. arginini* and *M. leonis* are identical. Disc electrophoresis patterns can thus be used as "finger prints" for the identification and classification of *Mycoplasma* strains[176,180].

The disc-electrophoretic technique was successfully applied in our laboratory for grouping and identifying large numbers of *Achromobacter*, *Alcaligenes*[120] and *Zymomonas* strains[201].

The number of reports on protein disc electropherograms of phyto-pathogenic bacteria is increasing considerably. Several representative strains of the following genera have been examined: *Agrobacterium*[105,114]. *Xantho-monas*[90], *Pseudomonas*[168] and *Erwinia*[193]. Polyacrylamide-gel protein patterns of over 250 *Agrobacterium* strains[120] were obtained under standardized conditions. They could be grouped in exactly the same three main and eight minor races, as obtained by DNA hybridizations (see p. 9). The homogeneity of cluster 2, containing *A. rhizogenes* and atypical *A. tumefaciens* strains was confirmed. The B6 and TT111 groups of cluster 1 form a heterogeneous population, containing *A. tumefaciens* and *A. radiobacter* strains. The correlation between protein profiles and DNA homology of these *Agrobacterium* strains is excellent. Our results suggest that the electropherogram technique, combined with DNA homology determinations and the analysis of primary structures of low molecular weight proteins can clarify the phylogenetic relationships among these phytopathogenic bacteria.

(ii) Zymograms

The zymogram technique consists of zone electrophoresis of enzymes in a suitable supporting medium, followed by staining *in situ* for specific enzyme activity[106]. This method is much more specific than the protein-pattern method and permits the rapid detection of minor structural differences among enzymes[189]. As explained above, identical electro-phoretic mobility of enzymes from different bacteria does not necessarily imply that the primary structures of these enzymes are identical.

Because multiple forms of esterases are easily detectable, they have been used for comparative zymogram analysis in a large number of bacteria, *e.g. Mycobacterium*[43], strains of *Bacillus thuringiensis*[163,164], lactic acid bacteria[156], several streptococci[138], coryneform bacteria[178], *Rhizobium*[158], *Vibrio*[48], *Leptospira*[92], and several bacteria from deep-sea sediments[99]. The widespread use of esterases for taxonomic and comparative purposes is open to criticism because esterase activity has been attributed to various enzymes hydrolyzing ester bonds with low specificity[121, 141], and non-enzymatic esterase activity has been reported[71]. Nevertheless, esterase patterns were generally used with good results as identification markers for several strains belonging to the genera mentioned earlier. It was found, for example, that most strains of *Rhizobium japonicum* lack esterase activity and that *R. trifolii* and *R. leguminosarum* show great similarity in esterase patterns whereas *R. meliloti* strains form a clearly distinct group.

Frequently, similar species-specific patterns were found in other bacterial enzymes such as hydrogenases[1], glucose-6-phosphate and glyceraldehyde-3-phosphate dehydrogenases in streptococci[226], lactic acid dehydrogenases in lactobacilli[86] and enteric bacteria[66]. Considerable intraspecific variation was observed for glucose-6-phosphate dehydrogenases and 6-phosphogluconate dehydrogenase from several *E. coli* strains[28].

Comparative zymogram analysis of one or two enzymes, as described above, has certain diagnostic and taxonomic merits when related bacterial species are investigated. It seems, impossible, however, to construct phylogenetic schemes of bacterial species on the basis of data derived from similarities or differences in the migration rates of enzymes. The magnitude of the differences in migration rates is not necessarily directly related to differences in primary structure (see above). Baptist, Shaw and Mandel[18] partly overcame these difficulties by determining electrophoretic mobilities of 8 enzymes in extracts of 27 *Enterobacteriaceae* (representatives of the genera *Escherichia*, *Shigella*, *Salmonella*, *Klebsiella*, *Serratia* and *Proteus*). The following enzymes were compared: malate, isocitrate, 6-phosphogluconate, glucose-6-phosphate and glutamate dehydrogenases, tetrazolium oxidase, phosphoglucomutase and catalase. The results were summarized in a matrix showing the dissimilarity coefficients of relative electrophoretic migrations. The less bacteria were related in terms of DNA base composition and overall similarity, the more their zymograms differed. The authors conclude that "similar electrophoretic migration rates of homologous proteins in the related organisms are more than fortuitous and probably reflected close homologies or even "identities" in the amino acid sequences of these enzymes." When performed in a similar manner, comparisons of several homologous enzymes by the zymogram technique may prove useful in unravelling the evolution of micro-organisms at the protein level.

ACKNOWLEDGEMENTS

One of us (J. de L.) is indebted to the Fonds voor Kollektief Fundamenteel Onderzoek and the Nationaal Fonds voor Wetenschappelijk Onderzoek (Belgium) for personnel and research grants.

References p. 72

REFERENCES

1   B. A. C. Ackrell, R. R. Asato and H. F. Mower, *J. Bacteriol.*, 92 (1966) 828.
2   J. M. Akagi, *J. Biol. Chem.*, 242 (1967) 2478.
3   R. P. Ambler, *Biochem. J.*, 89 (1963) 349.
4   R. P. Ambler, *FEBS Letters*, 18 (1971) 351.
5   R. P. Ambler, *Syst. Zool.*, (1974) in press.
6   R. P. Ambler and L. H. Brown, *Biochem. J.*, 104 (1967) 784.
7   R. P. Ambler, M. Bruschi and J. le Gall, *FEBS Letters*, 5 (1969) 115.
8   R. P. Ambler, M. Bruschi and J. le Gall, *FEBS Letters*, 18 (1973) 347.
9   R. P. Ambler and S. Murray, *Biochem. Soc. Trans.*, 1 (1973) 162.
10 R. P. Ambler and E. Taylor, *Biochem. Soc. Trans.*, 1 (1973) 166.
11 R. P. Ambler and M. Wynn, *Biochem. J.*, 131 (1973) 485.
12 J. R. Andreesen and G. Gottschalk, *Arch. Mikrobiol.*, 69 (1969) 160.
13 H. Bachmayer, A. M. Benson, K. T. Yasunobu, W. T. Garrard and H. R. Whiteley, *Biochemistry*, 7 (1968) 986.
14 H. Bachmayer, K. T. Yasunobu, J. L. Peel and S. Mayhew, *J. Biol. Chem.*, 243 (1968) 1022.
15 H. Bachmayer, K. T. Yasunobu and H. R. Whiteley, *Proc. Natl. Acad. Sci. (U.S.)*, 59 (1968) 1273.
16 H. Bachmayer, K. T. Yasunobu and H. R. Whiteley, *Biochem. Biophys. Res. Commun.*, 26 (1967) 435.
17 R. Bachofen and D. I. Arnon, *Biochim. Biophys. Acta*, 120 (1966) 259.
18 J. N. Baptist, C. R. Shaw and M. Mandel, *J. Bacteriol.*, 99 (1969) 180.
19 R. G. Bartsch, *Ann. Rev. Microbiol.*, 22 (1968) 181.
20 L. Baumann, P. Baumann, M. Mandel and R. D. Allen, *J. Bacteriol.*, 110 (1972) 402.
21 J. R. Benemann, D. C. Yoch, R. C. Valentine and D. I. Arnon, *Proc. Natl. Acad. Sci. (U.S.)*, 64 (1969) 1079.
22 A. M. Benson, H. F. Mower and K. T. Yasunobu, *Proc. Natl. Acad. Sci. (U.S.)*, 55 (1966) 1532.
23 A. M. Benson, K. Tomoda, J. Chang, G. Matsueda, E. T. Lode, M. J. Coon and K. T. Yasunobu, *Biochem. Biophys. Res. Commun.*, 42 (1971) 640.
24 A. M. Benson and K. T. Yasunobu, *J. Biol. Chem.*, 244 (1969) 955.
25 K. A. Bisset, in *12th Symp. Soc. Gen. Microbiol.*, Cambridge University Press, 1962, p. 361.
26 K. Bloch, in: V. Bryson and H. J. Vogel (Eds.), *Evolving Genes and Proteins*, Academic Press, New York, 1965, p. 53.
27 R. Bowen, *Paleotemperature Analysis*, Elsevier, Amsterdam, 1966, p. 265.
28 J. E. Bowman, R. R. Brubaker, H. Frischer and P. E. Carson, *J. Bacteriol.*, 94 (1967) 544.
29 D. J. Brenner, G. R. Fanning, K. E. Johnson, R. V. Citarella and S. Falkow, *J. Bacteriol.*, 98 (1969) 637.
30 D. J. Brenner, G. R. Fanning, F. J. Skerman and S. Falkow, *J. Bacteriol.*, 109 (1972) 953.
31 D. J. Brenner, G. R. Fanning and A. G. Steigerwalt, *J. Bacteriol.*, 110 (1972) 12.
32 D. J. Brenner, A. G. Steigerwalt, G. V. Miklos and G. R. Fanning, *Intern. J. Syst. Bacteriol.*, 23 (1973) 205.
33 M. Bruschi and J. le Gall, *Biochim. Biophys. Acta*, 263 (1972) 279.
34 M. Bruschi and J. le Gall, *Biochim. Biophys. Acta*, 271 (1972) 48.
35 M. Bruschi, J. le Gall and K. Dus, *Biochem. Biophys. Res. Commun.*, 38 (1970) 607.
36 M. Bruschi-Heriaud and J. le Gall, *Bull. Soc. Chim. Biol.*, 49 (1967) 753.

37 B. B. Buchanan and D. I. Arnon, *Adv. Enzymol.*, 33 (1970) 119.
38 B. B. Buchanan, R. Bachofen and D. I. Arnon, *Proc. Natl. Acad. Sci. (U.S.)*, 52 (1964) 839.
39 B. B. Buchanan and M. C. W. Evans, *Proc. Natl. Acad. Sci. (U.S.)*, 54 (1965) 1212.
40 B. B. Buchanan, H. Matsubara and M. C. W. Evans, *Biochim. Biophys. Acta*, 189 (1969) 46.
41 B. B. Buchanan and J. C. Rabinowitz, *J. Bacteriol.*, 88 (1964) 806.
42 M. Calvin, *Chemical Evolution*, Clarendon, Oxford, 1969, p. 278.
43 D. C. Cann and M. E. Willox, *J. Appl. Bacteriol.*, 28 (1965) 165.
44 C. W. Carter, S. T. Freer, Ng. H. Xuong, R. A. Alden and J. Kraut, *Cold Spring Harbour Symp. Quant. Biol.*, 36 (1971) 381.
45 P. E. Cloud Jr., in E. T. Drake (Ed.), *Evolution and Environment*, Yale University Press, New Haven, 1968, p. 1.
46 S. S. Cohen, *Am. Scientist*, 58 (1970) 281.
47 S. S. Cohen, *Am. Scientist*, 61 (1973) 437.
48 R. R. Colwell, V. I. Adeyemo and H. Kirtland, *J. Appl. Bacteriol.*, 31 (1968) 323.
49 P. Cuatrecasas, S. Fuchs and C. B. Anfinsen, *J. Biol. Chem.*, 243 (1968) 4787.
50 M. A. Cusanovich and D. E. Edmondson, *Biochem. Biophys. Res. Commun.*, 45 (1971) 327.
51 C. L. Cusumano, H. Taniuchi and C. B. Anfinsen, *J. Biol. Chem.*, 243 (1968) 4769.
51a M. O. Dayhoff, *Atlas of Protein Sequence and Structure*, Vol. 4, Natl. Biomed. Res. Found., Silver Spring, Md., 1969.
52 M. O. Dayhoff, *Atlas of Protein Sequence and Structure*, Vol. 5, Natl. Biomed. Res. Found., Silver Spring, Md., 1972.
53 K. Decker, K. Jungermann and R. K. Thauer, *Angew. Chem.*, 82 (1970) 153.
54 H. De Klerk and M. D. Kamen, *Biochim. Biophys. Acta*, 112 (1966) 175.
55 J. de Ley, *J. Appl. Bacteriol.*, 23 (1960) 400.
56 J. de Ley, in G. C. Ainsworth and P. H. A. Sneath (Eds.), *Microbial Classification*, 12th Symp. Soc. Gen. Microbiol., Cambridge University Press, 1962, p. 164.
57 J. de Ley, in Th. Dobzhansky, M. K. Hecht and W. C. Steere (Eds.), *Evolutionary Biology*, Meredith, 1968, p. 103.
58 J. de Ley, *Proc. Third Intern. Conf. Plant Path. Bacteria*, Pudoc, Wageningen, 197 (1972) 251.
59 J. de Ley, M. Bernaerts, A. Rassel and J. Guilmot, *J. Gen. Microbiol.*, 43 (1966) 7.
60 J. de Ley, J. de Smet, *et al.*, unpublished results.
61 J. De Ley, K. Kersters, J. Khan-Matsubara and J. Shewan, *Antonie van Leeuwenhoek*, 36 (1970) 193.
62 J. de Ley, K. Kersters, A. Reynaerts, H. Cattoir and R. Tytgat, unpublished results.
63 J. de Ley, I. W. Park, R. Tytgat and J. van Ermengem, *J. Gen. Microbiol.*, 42 (1966) 43.
64 J. de Ley, R. Tytgat, J. de Smedt and M. Michiels, *J. Gen. Microbiol.*, 78 (1973) 291.
65 D. V. Dervartanian, Y. I. Shethna and H. Beinert, *Biochim. Biophys. Acta*, 194 (1969) 548.
66 F. Detter and W. Rapp, *Zentr. Bakteriol. Parasitenk., Abt. I, Orig.*, 210 (1969) 220.
67 T. Devanathan, J. M. Akagi, R. T. Hersh and R. Himes, *J. Biol. Chem.*, 244 (1969) 2846.
68 R. E. Dickerson, *Sci. Am.*, 226 (4) (1972) 58.
69 R. E. Dickerson, *J. Mol. Biol.*, 57 (1971) 1.
70 R. E. Dickerson, T. Takano, D. Eisenberg, O. B. Kallai, L. Samson, A. Cooper and E. Margoliash, *J. Biol. Chem.*, 246 (1971) 1511.
71 W. K. Downey and P. Andrews, *Biochem. J.*, 96 (1965) 21c.
72 M. Dubourdieu and J. le Gall, *Biochem. Biophys. Res. Commun.*, 38 (1970) 965.
73 K. Dus, R. G. Bartsch and M. D. Kamen, *J. Biol. Chem.*, 237 (1962) 3083.

74   K. Dus, H. de Klerk, K. Sletten and R. G. Bartsch, *Biochim. Biophys. Acta*, 140 (1967) 291.
75   K. Dus, K. Sletten and M. D. Kamen, *J. Biol. Chem.*, 243 (1968) 5507.
76   K. Dus, S. Tedro, R. G. Bartsch and M. D. Kamen, *Biochem. Biophys. Res. Commun.*, 43 (1971) 1239.
77   R. V. Eck and M. O. Dayhoff, *Science*, 152 (1966) 363.
78   D. E. Edmondson and G. Tollin, *Biochemistry*, 10 (1971) 124.
79   J. Erwin and K. Bloch, *Science*, 143 (1964) 1006.
80   M. C. W. Evans, B. B. Buchanan and D. I. Arnon, *Proc. Natl. Acad. Sci. (U.S.)*, 55 (1966) 928.
81   M. C. W. Evans, A. V. Lord and S. G. Reeves, cited in ref. 94.
82   W. M. Fitch and E. Margoliash, *Science*, 155 (1967) 279.
83   J. L. Fox and J. R. Brown, *Fed. Proc.*, 30 (1971) 1242.
84   E. D. Garber and J. W. Rippon, *Advan. Appl. Microbiol.*, 10 (1968) 137.
85   J. M. Gardner and C. I. Kado, *Intern. J. Syst. Bacteriol.*, 22 (1972) 201.
86   F. Gasser, *J. Gen. Microbiol.*, 62 (1970) 223.
87   J. M. Ghuysen, *Bacteriol. Rev.*, 32 (1968) 425.
88   J. M. Ghuysen, J. L. Strominger and D. J. Tipper, in M. Florkin and E. H. Stotz (Eds.), *Comprehensive Biochemistry*, Vol. 26 A, Elsevier, Amsterdam, 1968, p. 53.
89   F. Gibson and J. Pittard, *Bacteriol. Rev.*, 32 (1968) 465.
90   H. S. Gill and M. N. Khare, *Phytopathology*, 58 (1968) 1051.
91   B. H. Gray, C. F. Fowler, N. A. Nugent and R. C. Fuller, *Biochem. Biophys. Res. Commun.*, 47 (1972) 322.
92   S. S. Green and H. S. Goldberg, *J. Bacteriol.*, 93 (1967) 1739.
93   D. O. Hall, R. Cammack and K. K. Rao, *Nature*, 233 (1971) 136.
94   D. O. Hall, R. Cammack and K. K. Rao, *4th Intern. Conf. "Origin of Life"*, Barcelona, June 1973.
95   G. Heberlein, J. de Ley and R. Tytgat, *J. Bacteriol.*, 94 (1967) 116.
96   G. D. Hegeman and S. L. Rosenberg, *Ann. Rev. Microbiol.*, 24 (1970) 429.
97   J. R. Herriott, K. D. Watenpaugh, L. C. Sieker and L. H. Jensen, *J. Mol. Biol.*, 80 (1973) 423.
98   G. Hind and J. M. Olson, *Ann. Rev. Plant Physiol.*, 19 (1968) 249.
99   M. A. Hogan and R. R. Colwell, *J. Appl. Bacteriol.*, 32 (1969) 103.
100 A. Holmgren, *Europ. J. Biochem.*, 6 (1968) 475.
101 R. Holmquist, T. H. Jukes and S. Pangburn, *J. Mol. Biol.*, 78 (1973) 91.
102 T. Horio and M. D. Kamen, *Ann. Rev. Microbiol.*, 24 (1970) 399.
103 N. H. Horowitz, *Proc. Natl. Acad. Sci. (U.S.)*, 31 (1945) 153.
104 N. H. Horowitz, in: *Evolving Genes and Proteins*, Academic Press, New York, 1965.
105 D. Huisingh and R. D. Durbin, *Phytopathology*, 57 (1967) 922.
106 R. L. Hunter and C. L. Markert, *Science*, 125 (1957) 1294.
107 M. Ikawa, *Bacteriol. Rev.*, 31 (1967) 54.
108 H. Iwasaki and S. Shidara, *Plant Cell Physiol.*, 10 (1969) 291.
109 L. H. Jensen, L. C. Sieker, K. D. Watenpaugh, E. T. Adman and J. R. Herriott, *Biochem. Soc. Trans.*, 1 (1973) 27.
110 R. A. Jensen, D. S. Nasser and E. W. Nester, *J. Bacteriol.*, 94 (1967) 1582.
111 R. A. Jensen and S. L. Stenmark, *J. Bacteriol.*, 101 (1970) 763.
112 P. W. Johnson and E. Canale-Parole, *Arch. Mikrobiol.*, 89 (1973) 341.
113 M. D. Kamen and T. Horio, *Ann. Rev. Biochem.*, 39 (1970) 673.
114 P. J. Keane, A. Kerr and P. B. New, *Australian J. Biol. Sci.*, 23 (1970) 585.
115 J. Kelly and R. P. Ambler, *Biochem. Soc. Trans.*, 1 (1973) 164.
116 S. J. Kennel, T. E. Meyer and M. D. Kamen, cited in ref. 76.

117 S. J. Kennel, T. E. Meyer, M. D. Kamen and R. G. Bartsch, *Proc. Natl. Acad. Sci. (U.S.)*, 69 (1972) 3432.
118 K. Kersters and J. de Ley, *Antonie van Leeuwenhoek*, 34 (1968) 393.
119 K. Kersters, J. de Ley, P. H. A. Sneath and M. Sackin, *J. Gen. Microbiol.*, 78 (1973) 227.
120 K. Kersters and J. de Ley, to be published.
121 J. R. Kimmel and E. L. Smith, *J. Biol. Chem.*, 207 (1954) 515.
122 D. T. Kingsbury, *J. Bacteriol.*, 94 (1967) 870.
123 B. C. J. G. Knight, *Bacterial Nutrition*, H. M. Stationery Office, London, 1938, p. 182.
124 E. Knight and R. W. Hardy, *J. Biol. Chem.*, 241 (1966) 2752.
125 E. Knight and R. W. Hardy, *J. Biol. Chem.*, 242 (1967) 1370.
126 T. Kodama and S. Shidara, *J. Biochem. (Japan)*, 65 (1969) 351.
127 K. A. Kvenvolden, J. C. Lawless and C. Ponnamperuma, *Proc. Natl. Acad. Sci. (U.S.)*, 68 (1971) 486.
128 E. J. Laishley, J. Travis and H. D. Peck, *J. Bacteriol.*, 98 (1969) 302.
129 J. le Gall and N. Dragoni, *Biochem. Biophys. Res. Commun.*, 23 (1966) 145.
130 J. le Gall and E. C. Hatchikian, *Compt. Rend.*, 264 (1967) 2580.
131 J. le Gall, G. Mazza and N. Dragoni, *Biochim. Biophys. Acta*, 99 (1965) 385.
132 E. Leifson, *Atlas of Bacterial Flagellation*, Academic Press, New York, 1960, p. 171.
133 E. B. Lewis, *Cold Spring Harbor Symp. Quant. Biol.*, 16 (1951) 159.
134 W. Lovenberg, B. B. Buchanan and J. C. Rabinowitz, *J. Biol. Chem.*, 238 (1963) 3899.
135 W. Lovenberg and B. E. Sobel, *Proc. Natl. Acad. Sci. (U.S.)*, 54 (1965) 193.
136 W. Lovenberg and W. M. Williams, *Biochemistry*, 8 (1969) 141.
137 M. L. Ludwig, R. D. Andersen, S. G. Mayhew and V. Massey, *J. Biol. Chem.*, 244 (1969) 6047.
138 B. M. Lund, *J. Gen. Microbiol.*, 40 (1965) 413.
139 A. Lwoff, *L'Évolution Physiologique. Étude des Pertes de Fonctions chez les Micro-organismes*, Hermann, Paris, 1943, p. 308.
140 R. Malkin and J. C. Rabinowitz, *Ann. Rev. Biochem.*, 36 (1967) 113.
141 M. Mandel, *Ann. Rev. Microbiol.*, 23 (1969) 239.
142 E. Margoliash and E. L. Smith, in: V. Bryson and H. J. Vogel (Eds.), *Evolving Genes and Proteins*, Academic Press, New York, 1965, p. 221.
143 L. Margulis, *Origin of Eukaryotic Cells*, Yale University Press, New Haven, 1970.
144 H. Matsubara, R. M. Sasaki, D. K. Tsuchiya and M. C. W. Evans, *J. Biol. Chem.*, 245 (1970) 2121.
145 S. G. Mayhew, *Biochim. Biophys. Acta*, 235 (1971) 276.
146 S. G. Mayhew, G. P. Foust and V. Massey, *J. Biol. Chem.*, 244 (1969) 803.
147 S. G. Mayhew and V. Massey, *J. Biol. Chem.*, 244 (1969) 794.
148 K. McCarthy, *Ph. D. Thesis*, Georgetown Univ., Washington D.C., quoted by Herriott *et al.*, ref. 97.
149 P. J. McLaughlin and M. O. Dayhoff, *Science*, 168 (1970) 1469.
150 P. J. McLaughlin and M. O. Dayhoff, *J. Mol. Evolution*, 2 (1973) 99.
151 T. E. Meyer, *Ph.D. Thesis*, Univ. California, San Diego, U.S.A. (1970), quoted in ref. 113.
152 T. E. Meyer, R. G. Bartsch, M. A. Cusanovich and J. H. Mathewson, *Biochim. Biophys. Acta*, 153 (1968) 854.
153 T. E. Meyer, J. J. Sharp and R. G. Bartsch, *Biochim. Biophys. Acta*, 234 (1971) 266.
154 K. Miki and K. Okunuki, *J. Biochem. (Japan)*, 66 (1969) 831.
155 K. Miki and K. Okunuki, *J. Biochem. (Japan)*, 66 (1969) 845.
156 T. Morichi, M. E. Sharpe and B.Reiter, *J. Gen. Microbiol.*, 53 (1968) 405.
157 S. Morita, *Biochim. Biophys. Acta*, 153 (1968) 241.
158 P. M. Murphy and C. L. Masterson, *J. Gen. Microbiol.*, 61 (1970) 121.

159 S. B. Needleman and T. T. Blair, *Proc. Natl. Acad. Sci. (U.S.)*, 63 (1969) 1227.
160 D. J. Newman and J. R. Postgate, *Europ. J. Biochem.*, 7 (1968) 45.
161 B. Nitomporn, J. L. Dahl and J. L. Strominger, *J. Biol. Chem.*, 243 (1968) 773.
162 C. Nolan and E. Margoliash, *Ann. Rev. Biochem.*, 37 (1968) 727.
163 J. R. Norris, *J. Appl. Bacteriol.*, 27 (1964) 439.
164 J. R. Norris and H. D. Burges, *J. Insect. Pathol.*, 5 (1963) 460.
165 J. M. Olson, *Science*, 168 (1970) 438.
166 W. H. Orme-Johnson, *Ann. Rev. Biochem.*, 42 (1973) 159.
167 W. H. Orme-Johnson, *Biochem. Soc. Trans.*, 1 (1973) 30.
168 B. C. Palmer and H. R. Cameron, *Phytopathology*, 61 (1971) 984.
169 J. A. Peterson and M. J. Coon, *J. Biol. Chem.*, 243 (1968) 329.
170 J. F. Petit, E. Munoz and J. M. Ghuysen, *Biochemistry*, 5 (1966) 2764.
171 P. G. Porque, A. Baldesten and P. Reichard, *J. Biol. Chem.*, 245 (1970) 2363.
172 A. R. Prévot, A. Turpin and P. Kaiser, *Les Bactéries Anaérobies*, Dunod, Paris, 1967, p. 2188.
173 S. C. Rall, R. E. Bolinger and R. D. Cole, *Biochemistry*, 8 (1969) 2486.
174 K. K. Rao and H. Matsubara, *Biochem. Biophys. Res. Commun.*, 38 (1970) 500.
175 S. Razin, *J. Bacteriol.*, 96 (1968) 687.
176 S. Razin and S. Rottem, *J. Bacteriol.*, 94 (1967) 1807.
177 S. Razin, J. Valdesuso, R. H. Purcell and R. M. Chanock, *J. Bacteriol.*, 103 (1970) 702.
178 K. Robinson, *J. Appl. Bacteriol.*, 29 (1966) 179.
179 R. F. Ross and J. A. Karmon, *J. Bacteriol.*, 103 (1970) 707.
180 S. Rottem and S. Razin, *J. Bacteriol.*, 94 (1967) 359.
181 T. G. Sacks, H. Haas and S. Razin, *Israel J. Med. Sci.*, 5 (1969) 49.
182 L. Sagan, *J. Theoret. Biol.*, 14 (1967) 225.
183 F. R. Salemme, S. T. Freer, Ng. H. Xuong, R. A. Alden and J. Kraut, *J. Biol. Chem.*, 248 (1973) 3910.
184 R. M. Sasaki and H. Matsubara, *Biochem. Biophys. Res. Commun.*, 28 (1967) 467.
185 K. H. Schleifer and O. Kandler, *Bacteriol. Rev.*, 36 (1972) 407.
186 J. W. Schopf, E. S. Barghoorn, M. D. Maser and R. O. Gordon, *Science*, 149 (1965) 1365.
187 K. T. Shanmugam and D. I. Arnon, *Biochim. Biophys. Acta*, 256 (1972) 477.
188 K. T. Shanmugam, B. B. Buchanan and D. I. Arnon, *Biochim. Biophys. Acta*, 256 (1972) 477.
189 C. R. Shaw, *Science*, 149 (1965) 936.
190 Y. I. Shethna, N. A. Stombaugh and R. H. Burris, *Biochem. Biophys. Res. Commun.*, 42 (1971) 1108.
191 Y. I. Shethna, P. W. Wilson and H. Beinert, *Biochim. Biophys. Acta*, 113 (1966) 225.
192 R. Sirevåg and J. G. Ormerod, *Science*, 169 (1970) 186.
193 J. H. Smith and D. Powell, *Phytopathology*, 58 (1968) 972.
194 J. R. Sokatch, *Bacterial Physiology and Metabolism*, Academic Press, New York, 1968, p. 443.
195 J. R. Sokatch and I. C. Gunsalus, *J. Bacteriol.*, 73 (1957) 452.
196 R. Y. Stanier, in: J. G. Hawkes (Ed.), *Chemotaxonomy and Serotaxonomy*, Academic Press, New York, 1968, p. 201.
197 R. Y. Stanier, N. J. Palleroni and M. Doudoroff, *J. Gen. Microbiol.*, 43 (1966) 159.
198 R. Y. Stanier and C. B. Van Niel, *J. Bacteriol.*, 42 (1941) 437.
199 S. L. Streicher and R. C. Valentine, *Ann. Rev. Biochem.*, 42 (1973) 279.
200 R. T. Swank and R. H. Burris, *Biochim. Biophys. Acta*, 180 (1969) 473.
201 J. G. Swings and J. de Ley, to be published.
202 C. Sybesma, *Biochim. Biophys. Acta*, 172 (1969) 177.

203 K. T. Tagawa and D. I. Arnon, *Nature*, 195 (1962) 537.

204 M. Tanaka, M. Haniu, K. T. Yasunobu, S. G. Mayhew and V. Massey, *Biochem. Biophys. Res. Commun.*, 44 (1971) 886.

205 M. Tanaka, M. Haniu, G. Matsueda, K. T. Yasunobu, R. H. Himes, J. M. Akagi, E. M. Barnes and T. Devanathan, *J. Biol. Chem.*, 246 (1971) 3953.

206 M. Tanaka, M. Haniu, K. T. Yasunobu and T. Kimura, *J. Biol. Chem.*, 248 (1973) 1141.

207 M. Tanaka, T. Nakashima, A. M. Benson, H. F. Mower and K. T. Yasunobu, *Biochemistry*, 5 (1966) 1666.

208 H. Taniuchi, C. L. Cusumano, C. B. Anfinsen and J. L. Cone, *J. Biol. Chem.*, 243 (1968) 4775.

209 R. Timkovich and R. E. Dickerson, *J. Mol. Biol.*, 79 (1973) 39.

210 A. Tissieres, *Biochem. J.*, 64 (1956) 582.

211 J. Travis, D. J. Newman, J. le Gall and H. D. Peck, *Biochem. Biophys. Res. Commun.*, 45 (1971) 452.

212 R. L. Tsai, I. C. Gunsalus and K. Dus, *Biochem. Biophys. Res. Commun.*, 45 (1971) 1300.

213 J. N. Tsunoda, K. T. Yasunobu and H. R. Whiteley. *J. Biol. Chem.*, 243 (1968) 6262.

214 J. G. Tully, M. F. Barile, R. A. Del Giudice, T. R. Carski, D. Armstrong and S. Razin, *J. Bacteriol.*, 101 (1970) 346.

215 R. C. Valentine, W. J. Brill and R. D. Sagers, *Biochem. Biophys. Res. Commun.*, 12 (1963) 315.

216 J. van Beeumen, *Biosystems*, (1974) in press.

217 J. van Beeumen and R. P. Ambler, *Antonie van Leeuwenhoek*, 39 (1973) 355.

218 B. van Lin and H. Bothe, *Arch. Mikrobiol.*, 82 (1972) 155.

219 H. Vetter and J. Knappe, *Z. Physiol. Chem.*, 352 (1971) 433.

220 H. J. Vogel, in: V. Bryson and H. J. Vogel (Eds.), *Evolving Genes and Proteins*, Academic Press, New York, 1965, p. 25.

221 B. Weinstein, *Biochem. Biophys. Res. Commun.*, 35 (1969) 109.

222 P. D. J. Weitzman and P. Dunmore, *FEBS Letters*, 3 (1969) 265.

223 P. D. J. Weitzman and P. Dunmore, *Biochim. Biophys. Acta*, 171 (1969) 198.

224 P. D. J. Weitzman and D. Jones, *Nature*, 219 (1968) 270.

225 R. H. Whittaker, *Science*, 163 (1969) 150.

226 R. A. D. Williams and E. Bowden, *J. Gen. Microbiol.*, 50 (1968) 329.

227 T. Yamanaka, *Ann. Rept. Biol. Works. Fac. Sci. Osaka Univ.*, 14 (1966) 47.

228 T. Yamanaka, *Nature*, 213 (1967) 1183.

229 T. Yamanaka, *Advan. Biophys.*, 3 (1972) 227.

230 T. Yamanaka, S. Takenami and K. Okunuki, *Biochim. Biophys. Acta*, 180 (1969) 193.

231 T. Yamanaka, S. Takenami, K. Wada and K. Okunuki, *Biochim. Biophys. Acta*, 180 (1969) 196.

232 D. C. Yoch and D. I. Arnon, *J. Biol. Chem.*, 247 (1972) 4514.

233 D. C. Yoch, J. R. Benemann, R. C. Valentine and D. I. Arnon, *Proc. Natl. Acad. Sci. (U.S.)*, 64 (1969) 1404.

234 D. C. Yoch and R. C. Valentine, *J. Bacteriol.*, 110 (1972) 1211.

235 D. C. Yoch and R. C. Valentine, *Ann. Rev. Microbiol.*, 26 (1972) 139.

Chapter IV

Biochemical Evolution in Animals

MARCEL FLORKIN

Department of Biochemistry, University of Liège, (Belgium)

INTRODUCTION

The evolution of organisms is, from the genetic viewpoint, the result of divergences in the gene pool of populations (the spreading of originally rare molecular mutants into the species), such divergences being determined by the interaction of population size, migration rate and selection intensity. Mutations at the level of genes, which take place in the configuration of the sequencing molecules of nucleic acids are (as stated in Chapter I) the core of the diachronic molecular epigenesis, the base sequence of DNA representing the permanent form of genetic information. Again as stated in Chapter I, from the extensive view point of thermodynamics, generalized biochemical evolution must be considered in the light of information theory as a diachronic process of information (negentropy) increase. This is a result of random changes, *i.e.* of statistical causality resulting from molecular and supramolecular structures in the context of the autocatalytic process of the reproduction of the sequencing macromolecules in conditions of energy flux remote from equilibrium. When we consider restricted evolution (see Chapter I), viewing individual things in the context of relations between organisms and environment, and recognizing that the evolving entities are populations of organisms, we do realize, as did the pioneers of the biological application of information theory, and rightly, that we must, in considering biochemical evolution as a process and knowing that thermodynamics can teach us nothing about mechanisms, take into consideration the intensive properties of the biomolecules implied. We do realize that information (as understood in information theory) is modulated by an intensive property related to biological relevance. In other words, biochemical evolution is to be

References p. 220

[79]

formulated in terms of molecular biosemiotics. Chapter I describes the flux of information from the associative configuration of biosemes (biosyntagms) of nucleic acids to the polypeptide chains entering into the constitution of effectors of cellular functions and of the catalytic components of the catenary biosyntagms, the products of which are themselves bioseme carriers.

The notion of the unity of the nature of catenary biosyntagms has been developed in Chapters I and II of Volume 29A. It constitutes one of the most striking elements of comparative biochemistry and has even been considered (see Chapter I) an argument against the notion of biochemical evolution. On the one hand, the enzymes involved in a given catenary biosyntagm evolve by the emergence of biosemes involved in metabolic regulation and, on the other, the metabolites which are the intermediates in a pathway may become the starting point for lateral extensions, as well as the terminal point of the chain for terminal extensions. Little is known about the extraction, from the system of sequencing biosyntagms, of new aspects of catenary biosyntagms. The case of the lactose synthetase pathway reveals the participation of a commutation of an enzyme (lysozyme) involved in another pathway, the resulting biomolecule (α-lactalbumin) acting in a ligand-induced modification of the signified of still another enzyme, belonging to another catenary biosyntagm. While the diachronic epigenesis of sequencing and sequenced molecules results from a number of fluctuations in the sequences of nucleotides or of amino acids involved, the diachronic epigenesis of catenary biosyntagms results from lateral and terminal extensions on the preserved backbone and from changes in the configuration of the enzymes involved, these changes introducing new biosemes in metabolic regulation. While colinearity exists between cistrons and polypeptides, in the field of non-polynucleotidic and non-polypeptidic biomolecules the colinearity obtains between phylogeny and biosynthesis, the emergence of new bio-molecules involving lateral or terminal extensions on the catenary pathway at the level of one of the metabolites which are successively formed along the chain, which is not modified. This should not, of course, be taken as an absolute general principle, as it often happens that convergences take place at the level of the biosynthesis of a metabolite.

In this review we cannot cover all the aspects of biochemical evolution in animals. Instead, we shall offer examples of individualized processes of biochemical evolution in animals.

In the present chapter, each section will be concerned with a concept of molecular biosemiotics rather than with a zoological concept and attention will remain focussed on aspects of integration at the molecular level. This approach, which concerns biochemical evolution as a process, differs from that adopted in the collective volumes of *Chemical Zoology*[1] (edited by M. Florkin and B. T. Scheer), in which the phyla of animals are considered in turn and their specific chemical components and processes described.

I. ANIMAL EVOLUTION

A. SHORT SURVEY OF ANIMAL SYSTEMATICS AND PHYLOGENY[2-9]

Within the phylum Protozoa, zoologists agree on the existence of several well-defined subphyla: Ciliata *(e.g. Paramecium)*, Sporozoa *(e.g. Monocystis)* and Cnidosporidia. What remains controversial is the question whether the Mastigophora or Flagellata *(e.g. Euglena)* should be distinguished from Sarcodina or Rhizopoda *(e.g. Amoeba)*, or kept together with the latter in the Subphylum Rhizoflagellata. It is nevertheless agreed that we must look among Flagellata for the most primitive animal monocellular organisms. The oldest animal fossils known are from the Pre-Cambrian rocks (coelenterates, brachiopods, annelids and echinoderms). Other phyla appear with relative frequency in the rocks of the Cambrian (beginning 500–700 million years ago and lasting about 100 million years).

The first arthropod fossils appear in the Cambrian and are represented by the trilobites. Other phyla with major representation in the Cambrian are the Brachipoda, the Porifera, the Coelenterata, the Mollusca (mainly Gastropoda) the Annelida and the Echinodermata.

After the Cambrian period, the Ordovician age (a time of warm shallow seas and of volcanic activity) lasted about 60 million years in the course of which corals appear in the fossil record, as well as Protozoa, Bryozoa and the now extinct graptolites in addition to trilobites, brachiopods, molluscs and echinoderms. For the determination of affinities the non-adaptive characters are of greater value (see Chapter I, Vol. 29A). According to Hadzi's theory, the metazoans arose by the internal division and cellularisation of a multinucleate protistan; he regards the turbellarian Platyhelminthes as the most primitive metazoans, derived from multinucleate protistans. From the rhabdocoel turbellarians, and more specifically from those that had adopted a sessile habit, the Coelenterata arose (the Anthozoa being the most primitive of Coelenterata). According to Hadzi's theory the most primitive Metazoa were triploblastic (triblastic) organisms (the turbellarian Platyhelminthes).

This theory does not seem to have found favour with zoologists, most of whom continue to adhere to the traditional view that an ancestral flagellate underwent mitotic divisions not followed by cell separation, the result being a "colonial" formation from which the diploblastic Metazoa (Porifera, Cnidaria, Ctenaria) were derived. Metazoa other than the diploblastic ones are characterized by their triblastic structure; a decisive step, as a third embryonic layer, mesoderm, then contributes to morphogenesis. The evolution of triblastic Metazoa (other than Porifera, Cnidaria and Ctenaria) is believed to have run along two main lines, the protostomians (annelid superphylum) and the deuterostomians (echinoderm superphylum). The protostomian line is relatively well-defined by the spiral structure of the egg segmentation, by the almost constantly teloblastic origin of the mesoderm, by the formation of the mouth at the level of the blastopore, and by the ventral position of the nervous system. The forms belonging to the annelid superphylum have in common a number of non-adaptive characteristics. When they have a larva, it generally belongs to the trochophore (or trochosphere) type (more or less spherical, bilateral symmetry and an equatorial band of cilia). Their coelom is schizocoelic, *e.g.* a blind cavity surrounded by mesoderm derived from a split in the embryonic mesoderm. Their eggs show vertical cleavage. Platyhelminthes, Annelida, Mollusca and Arthropoda belong to this superphylum.

When the forms belonging to the echinoderm superphylum have a larva, it is generally the pluteus type (ciliated ring around the mouth instead of equatorial as in the trochophore larva). The cleavage of the egg is radial and the coelom is enterocoelic, *e.g.* derived from pouches of the endoderm, each of which encloses a portion of the archenteric cavity. Echinoderms and chordates belong to this superphylum. The two main stocks appear to have diverged very early. Among protostomians, the mesoderm is either massive (acoelomates), distributed in separate masses in a cavity corresponding to the primitive blastocoele (pseudocoelomates) or else coelomic cavities (true coelomates) may be recognized in it. With the exception of the reproductive system, the acoelomates present a simpler organization than most other protostomians. Two phyla are distinguished among acoelomates (Platyhelminthes and nemertine worms). Platyhelminthes are characterized by the absence of an anus and a circulatory system. Nemertines have both.

Pseudocoelomates are characterized by the absence of a peritoneum lining the general cavity. Nematodes belong to them, as well as a number of other groups. Coelomate protostomians are divided into a number of groups. Among these a special position is given to the phylum to which the phoronidians, the Bryozoa and the Brachiopoda belong, all of which possess a lophophore, show a schizocoelian formation of the mesoderm and peculiarities in metamorphosis. The other coelomates show more homogeneity, *e.g.* with regard to formation of coelomic cavities from one or two coeloblasts, the organization of the trochophore larva, and the cell lineages which can be identified during the formation of the blastula. Metamerization is not observed in Sipuncula and is abortive in Echiuria. These two classes of marine vermiform animals probably descend from preannelid ancestors. The phylum Annelida constitutes the most remarkable use of the metameric organization among coelomates. The class Polychaetes is at the origin of Oligochaetes from which the hirudineans (Achaetes) are directly derived.

The molluscs are, from the embryological viewpoint, the most closely related to the annelids, the stages up to the formation of a trochophore larva being very similar. The molluscs differ from the annelids by the presence of a haemocoelic body cavity (*e.g.* an enlarged blastocoel filled with blood) and the absence of a coelom. They have abandoned metamerization. The largest class of molluscs is that of Gastropoda, showing, by way of a major evolutionary development, a tendency to torsion. They move by crawling on a ventral foot and are generally grazing feeders rasping food with the help of a radula. They generally have a dorsal shell. The Bivalvia are filter feeders and sedentary ; they have lost head, radula and ventral foot while their highly developed gills have become very efficient filter devices. The third and most developed class, the Cephalopoda, are carnivorous and free-swimming with shell reduced or lost, and with highly developed sense organs. The origin of molluscs can be traced to the calcification of a turbellarian ancestor.

Contrary to Mollusca, the Arthropoda have kept the metamery of their annelid ancestors but they have lost the primitive coelom (replaced by a hemocoele). Furthermore, they have acquired jointed appendages and developed a more or less rigid exoskeleton, derived from the cuticule.

The adaptive radiation of Arthropoda is unparalleled in the animal kingdom and reached its highest expression in the Insecta, who have overcome all varieties of media.

If the lineage of protostomian coelomates is well-defined and recognized by the majority of zoologists, the lineage of deutorostomian coelomates is more controversial and the kinship of echinoderms and chordates not universally accepted. Echinoderms are characterized by a radial symmetry (often pentamerous) and protrudable tube-feet. Members of the phylum possess a coelom which develops enterocoelically.

The Crinoidea are the most primitive echinoderms. They are sessile animals, with a stem bearing a theca from which the arms diverge. Asteroidea and Ophiuroidea are closely related and probably descend from common ancestors. The echinoids and the holothuroids have probably separated from the main line of free-living echinoderms between asteroids and ophiuroids.

Chordates share with echinoderms the characteristic of a coelom developing enterocoelically. The Protochordata may also be recognized as invertebrate chordates as they present at one stage or another the characteristics of chordates but without the vertebral column. The phylum Protochordata may be divided into three subphyla: Hemichordata Urochordata, Cephalochordata. Hemichordata are divided into enteropneusts (*e.g. Balanoglossus*; worm-like, free-living, filter-feeding, tornaria-pelagic larva) and Pterobranchiata (*e.g. Cephalodiscus*, small, tube-dwelling forms).

The Urochordata or tunicates present a notochord and nerve chord in the larval stage only *(e.g. Ciona)*. A tunic of a cellulose-like substance encloses the body. Ascidian larvae present typical chordate characteristics which are not found in adults.

The second main group of urochordates, Appendicularia *(e.g. Oikopleura)* lives a completely pelagic life. The third class of the subphylum urochordates (tunicates), the Thaliacea, are free-swimming pelagic forms. Cephalochordates, the third subphylum of the phylum Protochordata, include *Amphioxus* and have a full-length notochord in either the larval or the adult stages (but no vertebral column). From echinoderms, evolution leads to chordates *via* protochordates.

As far as vertebrates are concerned the fossil record begins in the Ordovician (fish scales in fresh water sediments). Ostracoderms (heavy armour plating) are among the fossils abundant in late Silurian and early Devonian. Ostracoderms are jawless, have no paired fins and only two semi-circular canals. They were bottom-living forms.

Reduction of the bony armour plating, with a tendency to free-swimming appears in a group of ostracoderms *(e.g. Birkenia)*. Ostracoderms became extinct by the end of the Devonian. Their only surviving relatives are the lampreys and hagfishes.

Gnathostomes are jawed vertebrates. While the early vertebrates probably had a large number of gill slits (each supported by a branchial arch of bone or cartilage), and while some ostracoderms had ten gills, these were reduced to seven in the immediate gnathostome ancestors.

The first of the remaining gill arches, losing its function of providing gill support "hinged forward to form jaws" (Price[2]). The first gnathostomes, now extinct, were the Acanthodia and later the placoderms (these appear in Upper Silurian. flourish through Devonian, and are extinct in Early Permian). They possessed a bony skeleton, paired fins and jaws (invention of internal bony scaffolding). The present-day jawed fishes, the Chondroichthyes and Osteichthyes, are on a line which passed through ostracoderm and placoderm stages, though they are not the descendants of the ostracoderms and placoderms.

Cartilagenous fishes, or Chondroichthyes (sharks, rays, dogfishes), have internal fertilization, separate external gill openings, a heterocercal tail and no swimbladder. The most ancient fossils of cartilagenous fishes appear in the Upper Devonian. From the four lines of their evolution which have been identified, two are extinct. One became the modern sharks and dogfishes, while the fourth one became the chimaeras. Bone appeared very early in the evolution of vertebrates (Early Devonian) and the bony fishes are known in more ancient sediments than the shark ancestors (Upper Devonian). Bony fishes (the class Osteoichthyes) have a bony skeleton, bony scales, an operculum covering gill slits and a swimbladder or lung.

Beyond the Devonian they are divided up in three sub-classes, Actinopterygii, Crossopterygii and Dipnoans. In Carboniferous times the Actinopterygii outnumbered the two other classes. The first stage in actinopterygian evolution is represented by the super-order Chondrostei, with incompletely ossified vertebrae and a functional lung. They reached a dominant position in Late Carboniferous and in Permian times. Two orders radiated, some forms of which are still surviving today: the sturgeons (*Acipenser*) on the one hand and the African genera *Polypterus* and *Calamoichthys* on the other. The second stage, the super-order Holostei, presents a reduction of the upper lobe of the heterocercal tail, more ossified vertebrae and conversion of the lung to a swimbladder. They reached their evolutionary peak in the Jurassic and Early Cretaceous periods to decline and leave only two North-American fresh-water survivors, the gar-pike *Lepidosteus* and the bow-fin *Amia*. Holostei were replaced by the super-order Teleostei. They continued the evolutionary trends already observed in the Holostei. Teleostei acquired a completely ossified skeleton, the tail became symmetrical (homocercal), the scales thinner and rounder. They met with considerable evolutionary success in all kinds of aquatic media.

References p. 220

The dipnoans have one or two rudimentary lungs, pseudochoanes and paired fins. Adapted to live in marshes subject to desiccation, they flourished at the end of the Primary. A few species subsist nowadays, disseminated on the three continents of the southern hemisphere. The dipnoans are not crossopterygii but they present a convergence with them with respect to certain characteristics. The Crossopterygii, the ancestral origin of tetrapod vertebrates, show a remarkable similarity to Amphibia with respect to skull bones, teeth (labyrinthodonts, with a highly folded structure in cross-section) and limbs. During Devonian times, the Crossopterygii evolved in two suborders, the Rhipidista and the Coelacanthini (to which the coelacanth Latimeria belongs). The earliest amphibians which appeared in Late Devonian, the ichthyostegids, resemble Rhipidista with respect to skull bones, teeth and the primitive structure of their vertebrae. However, they have an otic notch at each rear portion of the skull. This is related to an organ for the amplification of air vibrations, which is an adaptation to life on land.

In the course of the Carboniferous period, the primitive ichthyostegids evolved along several now extinct lines.

From the labyrinthodonts, though greatly removed, it has been possible to trace through a number of fossils the descent of the anurans (frogs and toads), while from the early ichthyostegid pattern the lepospondyls (with spool-like vertebrae) have diverged and led to the urodeles (newts and salamanders) and to apodans (blind, burrowing forms), a concept based on vertebral structure.

One of the impediments inherent in amphibian nature was the necessity of returning to water for reproductive purposes on the one hand and, on the other, the danger of dehydration resulting from loss of water through the skin.

The reptiles, still cold-blooded as were their amphibian ancestors, have kept the ossified dermic plates of their ichthyostegid ancestors but they have liberated themselves from the aquatic medium by the acquisition of the amniote egg which allowed reproduction without return to water. Large supplies of yolk allowed the young reptile to emerge from the egg as a fairly large and active creature. Along with this development went the elimination of the gill-breathing larval stage: reptiles emerged from the eggs as young adults without any further metamorphosis. A corollary is internal fertilization, which must occur before the relatively impermeable shell is deposited round the egg. The loss of water through the skin is greatly reduced by the presence of the scales.

The classification of reptiles is based on details concerning holes in the skull and is rather empirical. The subclass Anapsida is characterized by five skull perforations. It includes cotylosaurs (with cup-shaped vertebrae) and Chelonia (tortoises, turtles, terrapins). The subclass Synapsida has an extra opening at each side of the skull. It includes Pelycosauria (large belly-walker reptiles of the lower Permian, some presenting long neural spines), and their descendants the Therapsida (from late Permian to end of Triassic, more efficient locomotion than Pelycosauria as a consequence of the development of the limbs), the ancestors of mammals. Other subclasses of reptiles are the Parapsida, the Euryapsida (secondary adaptation to aquatic life) and the Diapsida, one order of which, Pterosauria or aerial Diapsida, flourished in the Jurassic and Cretaceous and contained arboreal forms with wing membranes supported by an elongated fourth digit, the fifth being lost.

Bird fossils appear in Jurassic times. They result in a second development of reptiles with respect to flight. The birds show a modified form of the fore-limbs as flying devices, with the hind limbs left for running and other activities; the scales modified into feathers are used not only in flight but as an adjunct to homeothermy. By the Cretaceous they had accomplished extensive radiation into the still extant orders.

As stated above, mammals derived from reptiles of the subclass Synapsida. It appears that several groups of this subclass (therocephalian, cynodont or ictidosaur reptiles) may have crossed the frontier to the mammalian structure independently. If this is so, then it seems probable that the origin of mammals was polyphyletic. Mammals present a series of organizational features: a four-chambered heart, mammary glands in females, three ear ossicles, specialized dentition with two successive sets of teeth, etc.

Mainly on the basis of dental features, Mesozoic fossil mammals are classified in four orders: Triconodonta, Symmetrodonta, Multituberculata and Pantotheria. Pantotheria appear to have given rise to the marsupial and placental stocks (evidence based almost entirely on the type and number of teeth).

When, as mentioned by Swain at the end of Chapter II (Vol. 29A), ichthyosaurs, plesiosaurs, dinosaurs and pterosaurs died out at the end of the Mesozoic, they left room for successful radiation of both marsupials and placentals. At the end of the Cretaceous, Australia was separated from Asia, and North America from South America.

The marsupials became the most successful forms in Australia where they occupied all the niches, while in South America a balance was maintained between marsupials and placentals. Placentals dominated in other regions and when, at the end of the Tertiary, both Americas were joined again, placentals from the North invaded South America and replaced placentals as well as marsupials there, while in Australia marsupials remained dominant (on other aspects of mammalian radiation, see Price[2]).

B. DRAWING PHYLOGENETIC TREES FROM AMINO ACID SEQUENCES

To establish the topology of a phylogenetic tree of organisms by considering amino acid sequences in the case of an orthologous protein such as cytochrome c, the sequences which have come to light may be compared with the sequence in man by way of reference. This comparison leads us to consider that the sequence nearest to that of man is the sequence of the *Rhesus* monkey which differs from the human sequence only by three residues of amino acids (24, 29, 69). The model sequence (common ancestor) carries all the

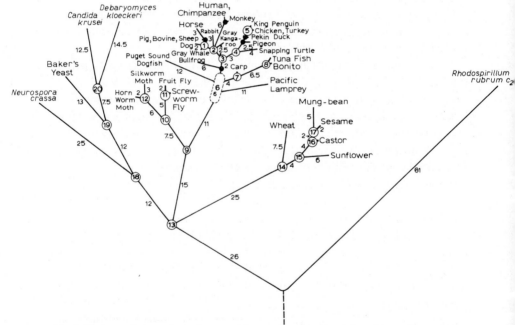

Fig. 1. Phylogenetic tree of cytochrome c. The topology has been inferred from the sequences. The numbers of inferred amino acid changes per 100 links are shown on the tree. The point of earliest time has been placed by assuming that, on the average, species change at the same rate.
(*Atlas*[10], Fig. 2–1, p. 8)

residues of amino acids common to the cytochromes of man and of the *Rhesus* monkey. The question that remains to be decided, at the level of the three residues concerned, is what protein presented the accepted point mutations. We see that, with respect to residues 24 and 29, the human protein presents the same character as do those of the other known mammals, while, with respect to residue 69, it is the *Rhesus* monkey who presents this character. It may be concluded that the common ancestor contained the three residues, Gln and Glu respectively, at 24 and 29, and Thr at site 69. The ancestral sequence is compared with the data known in the case of other species. Dayhoff, Park and McLaughlin (Chapter 2 of the *Atlas of Protein Sequence and Structure*[10], abbreviated in what follows as *Atlas*[10]) have devised a number of computer methods (matrix methods) in order to process the large amount of information involved in the construction of a phylogenetic tree from sequence data. The reader is referred to this presentation. The phylogenetic tree obtained for cytochrome *c* is reproduced in Fig. 1. The topology turns out to be the same as that which is based on morphological considerations. This applies also to the other phylogenetic trees derived from sequence data, which are found in the *Atlas*[10]. This is a gratifying and most important result which can be listed among the authentic successes of molecular biology. Though it does not prove that the nature of an organism is the sum of the natures of its constituent biomolecules and that this nature is the expression of the molecular order, it does lend support to the notion that the evolution of organisms and the molecular diachronic epigenesis (biochemical evolution) of their biomolecules are interrelated facets of the same natural reality.

C. ANIMAL PALEOBIOCHEMISTRY

Calvin[11] proposes to define as "molecular paleontology" the study of organic molecules which are left in rocks and which have been parts of molecular aggregates in prebiological evolution. This is, of course, a most interesting field which may become important in the future as and when criteria will be found to recognize abiogenesis or biogenesis, for instance in the case of mixtures of polyisoprenoids found in very old terrestrial sediments, or, eventually, in the case of organic molecules obtained from other planets, particularly Mars.

In this review we shall designate as animal paleobiochemistry[12,13] or molecular paleontology[14] the study of organic remains found in fossils (unearthed remains). Abelson, who discovered free amino acids in fossils, was the pioneer of paleobiochemistry. There is abundant literature on the subject. Abelson[15-18], Ezra and Cook[19], Ijiri and Fujiwara[20], Drozdova and Kochenov[21], Drozdova and Blokh[22], Armstrong and Tarlo[23], Oekonomidis[24] have detected free amino acids in fossil bones of teeth of various ages; Abelson[15], Fujiwara[25], Akiyama and Fujiwara[26], Hotta[27], Oekonomidis[24], Hare and Mitterer[28], Szoor[29], have detected free amino acids in fossil shells; Manskaya and Drozdova[30] in Ordovician graptolites; Armstrong and Tarlo[23] in Devonian conodonts. Chitin was detected in a Cambrian fossil by Carlisle[31]. Porphyrins have been identified in crocodile coprolites[32]. Polyhydroxyquinonoic pigments have been detected in Jurassic crinoids[33-35]. Olefins have been found in fossil shells of molluscs and brachiopods[37], and cellulose in fossil wood. Chitin is found in insect fossils of the Middle Eocene[38]. In Pleistocene bones, cholesterol esters, fatty acid esters and phospholipids are found, but not triglyceride[39], while in older fossils only fatty acids are found. In fossils, Everts et al.[40] have found that the fatty acids are essentially non-saturated, contrary to those found in modern bones (for more details, the reader is referred to Wyckoff[41]).

From the viewpoint of biochemical evolution, the organic constituents mentioned above are of limited interest with the exception of the just-mentioned difference in the degree of saturation of fatty acids. Sequenced molecules were believed, until 1961, not to be present in fossils older than approximately 100000 years. In 1960, Jones and Vallentyne[42] demonstrated the presence of conchiolin remains in a 100000-year-old *Mercenaria* shell.

In 1961, the outlook changed when Florkin et al.[36] demonstrated the preservation of shell proteins in fossils for much longer periods (tens of millions of years) and in one example, in a fossil of the Eocene (60 million years). After isolating particles of mother of pearl from the fossil shells and removing the remains of the prismatic structure, these particles were powdered in a mortar. The resulting powder was washed with boiling water until free amino acids had been completely removed and then treated with successive portions of 6 N HCl until the mineral constituents were fully decomposed (end of effervescence). The residue was dialyzed against running water and for 4 h against distilled water. It was evaporated to dryness and reflux-condensed in boiling 6 N HCl for 24 h, evaporated under reduced pressure, treated with active charcoal and brought to a known volume of which aliquots were used for amino acid determinations by chromatography on Dowex 50 according to the method developed by Moore and Stein. A mixture

TABLE I

"CONCHIOLIN" OF MOTHER-OF-PEARL
Amino acid composition: molecular fraction per 100 molecules of amino acids.
(Florkin et al.[36])

	Nautilus Eocene (60 million years)	Aturia Oligocene (40 million years)	Iridina Holocene (10000 years)	Nautilus Modern
Aspartic acid	8.7	9.0	10.1	9.0
Threonine	5.6	3.8	1.6	1.5
Serine	24.0	16.7	7.8	10.9
Glutamic acid	15.6	11.9	5.0	5.5
Proline	3.7	4.8	2.0	1.8
Glycine	20.8	23.3	29.8	35.7
Alanine	9.7	15.2	28.2	27.2
Valine	3.1	5.7	4.0	2.2
Isoleucine	3.1	3.3	2.4	1.8
Leucine	5.6	6.2	4.0	1.9
Tyrosine	0	0	0	2.4
Phenylalanine	0	0	0	2.4
Histidine	0	–	0	0
Lysine	0	–	2.3	0
Arginine	0	–	1.9	0

of amino acids was obtained whose pattern corresponded to that of nacre conchiolin (Table I, Fig. 2). The older the fossil, the larger the amount of fossil material necessary to obtain a quantity of nitrogen between 132 and 135 μg N. In spite of constancy of the amounts of nitrogen, the amounts of amino acids obtained were smaller in the older fossils, the difference being accounted for by the ammonia liberated in the course of hydrolysis. The older the fossil, the lower the proportion of glycine and alanine. It was also found that particles of the decalcified material showed positive biuret and amido-schwarz reaction. Fig. 3 shows the electron micrographs of nacreous organic remnants, after decalcification with chelating agents, of the same material used to obtain the data of Table I and of Fig. 2. These electron micrographs justify our discarding the possibility of a contamination of the material by parasites, epibiontes, etc. and they confirm the nature of the protein as belonging to mother-of-pearl

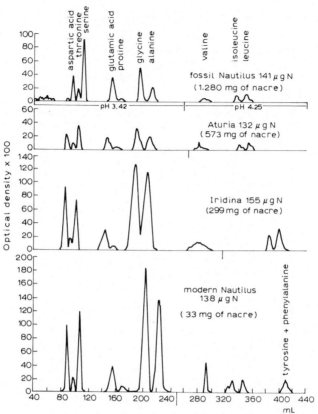

Fig. 2. Separation by chromatography on Dowex 50, using the Moore and Stein technique, of amino acids resulting from the hydrolysis of the residue of nacre after washing out the free amino acids, decalcification and dialysis, of modern *Nautilus*, fossil (Eocene) *Nautilus*, Recent (Holocene) *Iridina* and fossil (Oligocene) *Aturia*. The ultrastructure of the samples used is shown in Fig. 3. (Florkin *et al.*[36])

layer. The purpose of the analysis conducted by Florkin *et al.*[36] was to establish the presence of proteins in the fossil shell structures. To reach conclusions concerning evolution on the basis of amino acid content never was the authors' purpose. The material used (mother-of-pearl conchiolin) is composed of the water-soluble nacrine and the insoluble nacroine, a scleroprotein of the nature of a chitinoproteic complex. This is non-homogeneous material, the composition of the final material depending on the treatment adopted. It would be even more hazardous to speculate on the comparative data regarding amino acid composition obtained on whole shells, as it has been

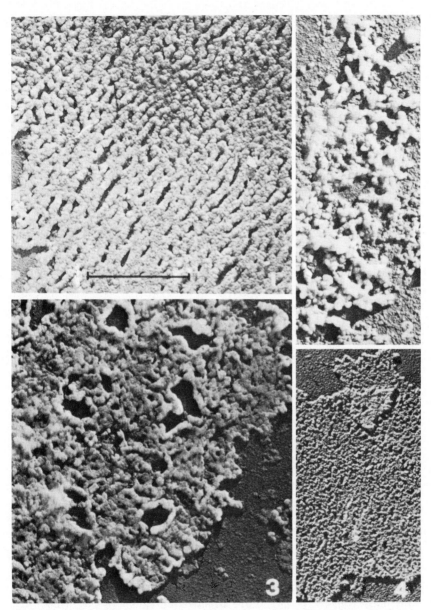

Fig. 3. Electron micrographs of nacreous organic remnants after decalcification with chelating agents. Fragments of lace-like reticulated sheets of conchiolin. (Florkin *et al.*[36]) (1) *Nautilus macromphalus* (Cephalopoda, Nautiloida). Recent (Nautiloid pattern). ×36000. (2) *Nautilus* sp. (Cephalopoda, Nautiloida). Eocene (60 million years). Nautiloid pattern still recognizable in some regions. × 36000. (3) *Aturia* sp. (Cephalopoda, Nautiloida). Oligocene (40 million years). Nautiloid pattern still recognizable. ×27000. (4) *Iridina spekii* (Pelecypoda, Mutelida). Holocene (10000 years). Pelecypod pattern. × 27000.

References p. 220

TABLE II

COMPARISON OF AMINO ACID COMPOSITIONS FOR "CONCHIOLINS" OF PRISMS FROM
MODERN AND FOSSIL SHELLS

(Bricteux-Grégoire et al.[43])

| | Modern | | | | | | Fossil | | | | | |
| | Atrina nigra | | | Pinna nobilis | | | Pinna affinis | | | Inoceramus | | |
	μg/g	per-cent N	mol. fr.	μg/g	per-cent N	mol. fr.	μg/g	per-cent N	mol. fr.	μg/g	per-cent N	mol. fr.
Lysine	145	0.7	0.8	110	0.5	0.6	tr			1.58	0.7	4.8
Histidine	70	0.3	0.4	89	0.4	0.5	tr			1.13	0.5	3.2
Arginine	252	1.1	1.1	365	1.5	1.7	tr			0.96	0.4	2.4
Aspartic acid	1750	9.6	10.4	3920	20.5	23.3	23.4	2.6	11.1	3.03	1.6	10.0
Threonine	173	1.1	1.1	231	1.4	1.5	8.8	1.1	4.7	1.18	0.7	4.4
Serine	· 530	3.7	4.0	617	4.1	4.6	27.5	3.8	16.5	3.88	2.5	16.4
Glutamic acid	330	1.6	1.8	348	1.7	1.9	31.8	3.1	13.6	5.13	2.4	15.5
Proline	314	2.0	2.2	316	1.9	2.2	tr			tr		
Glycine	4390	42.4	46.2	3500	32.4	36.9	26.8	5.2	22.5	3.60	2.3	21.3
Alanine	530	4.3	4.7	640	5.0	5.7	20.0	3.3	14.1	1.62	1.2	8.1
Valine	930	5.8	6.3	833	4.9	5.6	9.2	1.1	5.0	1.30	0.8	4.9
Isoleucine	334	1.8	2.0	558	3.0	3.4	7.9	0.9	3.8	0.99	0.5	3.4
Leucine	1220	6.8	7.4	740	3.9	4.5	12.2	1.4	5.9	1.69	0.9	5.7
Tyrosine	2130	8.6	9.3	1200	4.6	5.3	tr			tr		
Phenylalanine	472	2.1	2.3	508	2.1	2.4	7.6	0.7	2.9	tr		
		91.9	100.0		87.9	100.1		23.2	100.1		15.5	100.1
Amino nitrogen	1930			2020			96			20		
Ammonia	70	3.0		90	3.7		131	113			14.1	56.7

shown that the prismatic layer and the mother-of-pearl layer contain conchiolins of different compositions.

Bricteux-Grégoire et al.[43] have isolated individual prisms from fossil shells, namely those of *Pinna affinis* (London Clay, Lower Eocene) and *Inoceramus* sp. (Gault, Cretaceous), and have decalcified these structures. Demineralization of the prisms, performed in saturated aqueous solutions of EDTA (ethylene-diamine tetraacetic acid, disodium salt or Titriplex III), and in 2-percent followed by 25-percent hydrochloric acid solution, leaves transparent, glassy, brittle, sometimes tubular fragments of substance, these being the remains of the prism sheaths. After being washed to remove free amino acids, the isolated sheaths from prisms isolated from both modern and fossil specimens have been hydrolyzed and the products analyzed. The results confirm the presence of proteins in the material (Table II).

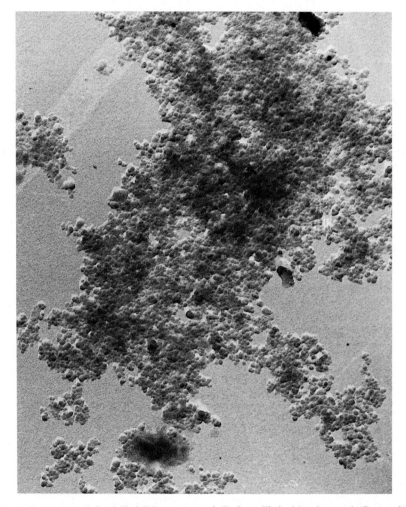

Fig. 4. Fragments of decalcified Dinosaur egg shells (very likely *Megalosaurus*). Corpuscles of the thin stratified sheets. × 42000. (Voss-Foucart[57])

After the persistence for long geological periods of proteins in fossil shells had been recognized in 1961, collagen was extensively studied in fossil bones in the years following 1963. It has been detected by X-ray diffraction by Isaacs et al.[44] in bones, or dentine of fossils, some of these going back to the Devonian; by electron microscopy by Little et al.[45] in buffalo bones; by Wyckoff et al.[46] in the calcaeneum of a Pleistocene Equus; by Shackleford and Wyckoff[47], and Wyckoff and Doberenz[48,49] in Miocene and Pleistocene bones and teeth; by Pawlicki et al.[50] in the bones of dinosaurs of the upper Cretaceous; by Doberenz and Lund[51] in a lower Cretaceous fossil and by Doberenz and Wyckoff[52] in Pleistocene bones.

Fossil collagen composition was first studied by Wyckoff et al.[46] who obtained a series of amino acids, including hydroxyproline, in the hydrolysis products of decalcification residues of Pleistocene mammalian bones. Ho[53-55] compared the composition of fossil and recent collagens and observed that fossil collagen generally contains less leucine, phenylalanine and tyrosine, and more glycine. Miller and Wyckoff[56] studied the proteins preserved in more ancient bones (bones and teeth of Jurassic and Cretaceous dinosaurs). As shown in the author's laboratory by Voss-Foucart[57], the slow decalcification of dinosaur egg shells (very likely threopods belonging to the genus Megalosaurus) of the upper Cretaceous of Provence, liberates two different structures: one composed of white ribbons forming an external and an internal net joined by thin hollow tubes which correspond to the air channels, the other consisting of thin brown sheets. Both structures still contain protein whose ultrastructure was recognized by electron microscopy (Figs. 4 and 5). The determination of their amino acid composition, after elimination of free amino acids by washing, desiccation and hydrolysis revealed proteins in both structures, and differences in their composition. In the author's laboratory, Foucart et al.[58] have carried out analyses of three different samples of fossils belonging to the order Graptolidea (Silurian Pristiograptus gotlandicus, Pristiograptus dubius; Silurian monograptidae, contaminated with a few Retiolitidae; Ordovician Climacograptus typicalis, provisional identification) and found in their hydrolysates amino acids of protein origin.

Preserved proteins have also been detected in fossil shells of Brachiopoda[59,60] from the Ordovician and Carboniferous. Carlisle[31] found protein remains in a Cambrian fossil of a pogonophore, Hyolithellus.

The persistence of proteinaceous residues in fossil shells of great age, revealed first in 1961 by Florkin et al.[36], has been confirmed repeatedly in

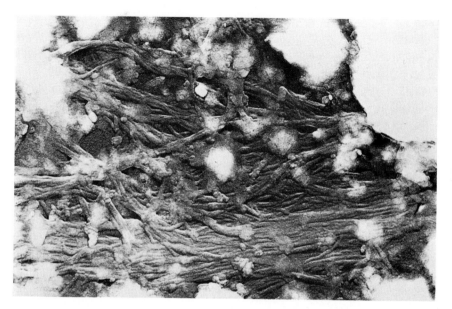

Fig. 5. Fragments of decalcified Dinosaur egg shells (very likely *Megalosaurus*). Fibres of the white ribbons forming the superficial networks and the wall of the air channels. Negative. × 48 000. (Voss-Foucart[57])

later studies from Wyckoff's laboratory and by other authors (literature in Wyckoff[41]). Since that time, the notion of the existence of preserved proteins in very ancient fossils has been critically discussed. It was first suggested that free amino acids, liberated from indigenous proteins in the fossil, may associate with mineral constituents in the form of carbamates and be accumulated locally. But it is clear that the demineralization of the fossil (decalcification by hydrochloric acid) would eliminate such compounds. Small peptides, on the other hand, are eliminated by the dialysis of the samples before hydrolysis.

Contamination may also introduce three different kinds of artefacts: (1) Organisms contemporary with the fossilized organism; (2) Contamination by substances surrounding the fossil; (3) Contamination during the manipulation and the analysis.

In order to eliminate the possibility of contamination by parasites, epibionts, etc., control by electron microscope is of interest. It is also necessary to handle the material with caution and avoid any bacterial or material contamination. Another point raised by Abelson[18] and by Mitterer[61] was

that recrystallisation of the inorganic material of shells would make preservation of proteins inconceivable. Grandjean et al.[62], and Hall and
Kennedy[63] have shown by X-ray diffraction that in the shells of the Tertiary
recrystallization of the aragonite into calcite is rare. It is more common in
secondary specimens (particularly of the Jurassic) and is generalized in
Primary shells. A large amount of literature shows that the recrystallization
replacing aragonite by calcite in shells has taken place in situ and that protein
remains persisted in recrystallized shells. As Hudson[64] writes:

"the conditions which favour the preservation of organic matrix are not the same as those which
favour the preservation of original carbonate mineralogy, especially in aragonitic forms".

The reviewer has already expressed the view that composition data, though
they do demonstrate the presence of protein remains should not serve as the
foundation for evolutionary theories because the material analyzed—e.g. in
the case of shells, and particularly of whole shells—is of a very heterogenous
nature.

There is another consideration which detracts from the value of composition data. It is the fact that proteins have in the course of geological time
been modified and that native proteins have been transformed into paleoproteins (paleization).

Attacking the problem experimentally, Grégoire[65] in the author's
laboratory exposed fragments of Modern shells to a few of the factors involved in paleization, such as heat and pressure. This study is now in progress.

It must be kept in mind that the study of amino acids, either free or polymerized, in fossils has been pursued in the context of the history of sediments.
Proteins may form a very small portion of sediment matter and so hold little
interest for the geochemists. An electron-microscopic study of shells may be
of interest in comparative submicroscopical morphology, independently of
the detection of either contaminations or of gross morphological changes
in proteins due to paleization.

Then again, for those who are interested in the biochemical evolution of
proteins and who have pursued the study of paleoproteins since their
discovery in 1961, it is the biochemical aspect which comes to the fore in the
context of the search for new data of importance to phylogeny. Fractures or
loss of terminal or lateral parts do not diminish the importance of protein
remains since the primary structure of paleoproteins (the only valuable
homology test) is unlikely to be modified in paleization except with respect

to the length of chains. Studies on amino acid sequences in fossil proteins should be very rewarding when we succeed in decomposing them into small peptides.

II. THE GENETIC LOSS OR REPRESSION OF BIOSEMES IN ANIMALS

One general feature of animal metabolism can be recognized even among animal flagellates and ciliate Protozoa. The ciliates, for instance, require ten amino acids in their diet for normal growth (arginine, histidine, isoleucine, leucine, lysine, methionine, phenylalanine, threonine, tryptophan, valine) (for the literature, see Vol. 1 of *Chemical Zoology*[1]). The same applies to fish such as the chinook salmon (Halver and Shanks[66]), reptiles (Coulson and Hernandez[67]), birds (Brown[68]) and the growing mammal (Scull and Rose[69]).

Nitrogen fixation from the air is of paramount importance to the economy of the living world but it takes place only in some microorganisms, mostly bacteria. Inorganic nitrogen compounds such as ammonia and nitrates can be used as major nitrogen sources by phytoflagellates and by plants; however, all Protozoa other than phytoflagellates require, in addition to Metazoa, some form of organic nitrogenous compound. This is an example of regressive biochemical evolution, whereby the enzymatic systems required to build organic compounds from inorganic ones are lost or repressed.

As is well known, vitamins present a very clear example of regressive biochemical evolution. It seems that in the first precellular structures which appeared in the transition from the abiogenic organic continuum to the biochemical continuum, the catalysis of organic biosynthesis was inorganic in nature. In progressive biochemical evolution, these simple systems became more efficient by the inclusion of reaction-specific proteins (enzymes) and of organic functional analogues of the inorganic elements (coenzymes). The higher plants can synthesize these coenzymes with the exception of vitamin B_{12}. In animals, the dependence on the biochemical continuum is much greater. Depending on the phylogenetic situation of the loss of the enzyme system of biosynthesis, the forms of distribution of the chemical needs in the sections of animal systematics are quite different and clearly related to the course of biochemical evolution. Vitamin A, for instance, is unknown in plants. It is acquired in animals by a biochemical conversion of certain carotenoids borrowed from plants, the ability to synthesize them having been lost or repressed in all animals.

References p. 220

The chemical needs of animals are generally satisfied by food components though in the case of ruminating cattle the rumen microorganisms provide thiamine, riboflavin, nicotinic acid, pyridoxine, pantothenic acid, biotin, folic acid, vitamin B_{12} and vitamin K in amounts sufficient to meet the entire demands of the adult animal.

The nitrogenous requirements of Protozoa have been extensively studied (for an extensive treatment of the subject, the reader is referred to Volume 1 of *Chemical Zoology*[1]). The nitrogenous requirements of Protozoa may be satisfied by simple and non-specific nitrogenous compounds such as ammonia or the amino groups of any kind of amino acids. In a number of examples vitamin B_{12}, thiamine and biotin are the only organic molecules which are not synthesized.

The animal flagellates (Zoomastigophorea) so far studied present an entirely different picture; here the pattern of enzymapheresis (enzyme deletion or repression) is surprisingly similar to that of Crustacea, insects or vertebrates. It is of interest to note that the soil amoeba, *Hartmanella ryzodes*, requires only seven out of the ten commonly essential amino acids: arginine, isoleucine, leucine, lysine, methionine, threonine and valine (Band[70]). Another soil amoeba, *Acanthamoeba*, has been shown (Adam[71]) not to require lysine or threonine but to include six amino acids in its list of essentials due to the fact that it needs phenylalanine. It may be the case that these forms (which do not synthesize thiamine or biotin, either)—so far as these characteristics are concerned—are on a regressive evolutionary pathway from algae towards the zooflagellate and animal types of limitation.

The ciliated Protozoa conform to the "animal pattern". All ciliated Protozoa have lost the enzymatic system for the construction of the purine ring. In addition they have lost the enzymes for the production of guanine nucleotides from inosinic acid (which is inosine monophosphate). As a consequence, guanine has become an absolute requirement. This has been shown in *Tetrahymena pyriformis*. The same regressive evolution has taken place in other ciliates such as *Paramecium aurelia*, *Paramecium multimicronucleatum* and *Paramecium caudatum* for which guanine is a dietary requirement. It cannot, however, be used as the free base but only in the form of its nucleoside or nucleotide, thus demonstrating that these species of *Paramecium* have lost the ribosylating activity for guanine as well (literature in Kidder[72]). The Trypanosomids also require purine but can use any one; consequently they have preserved the faculty (lost by Tetrahymenid Ciliates) of shifting around substituent groups on the purine ring.

III. AMINO ACIDS AND THEIR POLYMERS IN THE BIO-CHEMICAL EVOLUTION OF ANIMALS

A. INTRODUCTION

Polypeptides and proteins are sequenced biosyntagms reflecting the sequencing structure of the nucleic acids of genes by the double mechanism of the transcription of the DNAs into messenger RNAs and the translation of messenger RNAs into polypeptide chains. One aspect of this mechanism is the colinearity of cistrons and the polypeptide chains resulting from the flux of information starting at the level of the cistron.

The progress of molecular genetics and of molecular biology has led us to consider that, in the replication of the genetic material which governs the synthesis, during meiosis, of the polynucleotides of the gametes, fluctuations are bound to take place as the inevitable concomitant of the autocatalytic process. We also believe these fluctuations are followed in the course of the diachronic epigenesis of sequencing molecules by a multiplication, in the genome, of polynucleotides, accompanied by duplication and independent epigenesis, and considerable redundancy. It appears that only a part of the DNA molecules is involved in protein synthesis, a large proportion being involved in regulation at the genome level (see the *hypothetical* model of Britten and Davidson in section 17, Chapter I; p. 42).

At the level of sequenced macromolecules (polypeptides and proteins, including enzymes) the process of variation at the level of polynucleotides is expressed by point mutations, deletions, insertions, polymorphism, hybridisation and partial internal duplication (see Chapter I).

The signified of sequenced macromolecules may remain unchanged (orthologous homology) or it may be changed (commutations, paralogous homology). In this review, as stated in Chapter I, isology means chemical kinship. As in the writings of Redfield, Margoliash and Neurath, homology means genetic isology, *i.e.* a degree of isology greater than could occur by chance. The concept is expressed by Mrs. Dayhoff[10] as "relatedness". Redfield defines homology as "a generic relationship between configurations". Analogy is used here in the sense defined by Redfield as common biochemical activity*. Following Fitch[72a] homology may be orthologous or paralogous and, as proposed by Neurath, it may be intraspecial or inter-

---

* It may be noted here (see Chapter I) that in the terminology used by Anfinsen and Acher, homology means what we call analogy, and that Ingram calls homology what we call isology.

special (see Chapter I).

We can now express the diachronic epigenesis of the flux of information from polynucleotides to proteins in terms of protein lineages, by drawing the genealogic tree of homologous polypeptidic chains. The system revealed in this way is one of evolutionary descent. It need imply no change of signified (orthologous homology). It may in other cases imply radical changes in signified (commutations). In many cases protein diachronic epigenesis has been at the origin of the emergence of biosemes responsible for ligand-induced modulations of the signified and for feedback regulations of catenary biosyntagms. Commutations as well as heterotypic expressions are involved in the diversification of catenary pathways.

B. EVOLUTION OF PROTEIN SYNTHESIS IN ANIMALS

1. Introduction

The morphological and the physiological aspects of a pluricellular organism result from a regulated system of cell differentiations which accompanies the multiplication of cells in growth until a regulated volume of the constituent parts is reached.

While they maintain a common central metabolic catenary biosyntagm which may be thought of as frozen, each category of cells in an organism puts into operation a particular set of information contained in the genome. This information flows through the sequencing biosyntagms commanding the synthesis of a definite set of sequenced biosyntagms (proteins and peptides). If we consider, for instance, a mammalian organism with respect to the differentiation of its cells in the domain of proteins, we can distinguish a number of differentiated aspects.

Among the cells of ectodermal origin, the epidermic cells synthesize keratin; the eosinophilic cells of the adenohypophysis, the growth hormone and the lactogenic hormone; the basophilic cells of the adenohypophysis, the thyrotropic hormone, ACTH, the follicle-stimulating hormone, etc. Among the cells of endodermal origin, those of the salivary gland bio-synthesize ptyalin; the peptic cells of the gastric mucosa, pepsinogen; the pancreatic exocrine cells, a lipase, glyceridases, endopeptidases and nucleases; the pancreatic endocrine cells, insulin and glucagon; the cells of intestinal mucosa, secretin, cholecystokinin, pancreozymine, exopeptidases; the thyroid cells, thyroid hormones and calcitonin; the parathyroid cells, the parathyroid hormones. Among the cells of mesodermal origin, the cells of the suprarenal cortex and of the gonads synthesize the enzymes for the biosynthesis of steroid hormones.

None of these differentiations are found in the cells of an insect, whose common central metabolic catenary biosyntagm is nevertheless basically the same as that in mammals, for example. The differentiated aspects of protein biosynthesis, taken together, satisfy (part of) the definition of a zoological category when the organisms are considered at the molecular level of integration and can thus be considered under the aspect of diachronic molecular epigenesis.

2. Lineages of proteins and peptides

In Chapter I, the molecular diachronic epigenesis of animal cytochromes was reviewed. It is an example of the diachronic epigenesis of an orthologous protein, which at the molecular level has not changed in biological activity while in animal evolution many proteins show changes in the degree or nature of this activity. The latter phenomenon is often interpreted as the result of gene duplication and of progressive changes in biological activity resulting from the individualization of modified genes, whereas in orthologous molecules the changes appear as point mutations of the same gene. As stated above, a study of primary structure in orthologous proteins enables us to trace the evolution of these proteins along the phylogenetic tree of organisms and consequently to test the reliability of this tree as well as, hopefully, to clarify obscure points.

A collection of data on primary protein structures is contained in the *Atlas*[10]. While the *Atlas*[10] contains 267 sequences for vertebrates (236 of which are related to mammals), there are only 12 sequences of invertebrates (plus three in the supplement, 1973). The molluscs, for instance, are represented in the collection by eledoisin, one of the peptide toxins of *Octopus* venom. Plants are also poorly represented, with 17 sequences in all.

While polynucleotides (sequencing biosyntagms) carry the biosemes involved in the flux of information of the biosynthesis of polypeptides (sequenced biosyntagms), the latter appear as the effectors of synchronic epigenesis and are among the effectors of cellular physiology. They also act as catalysts in the successive steps of catenary metabolic biosyntagms, to which we shall return later.

The system of sequenced biosyntagms, expressing the generic relationships

TABLE III

SOME LINEAGES OF POLYPEPTIDIC CHAINS
(an aspect of the system of polypeptidic chains)
(literature in *Atlas*[10])

1. cytochrome c; cytochrome c_1, cytochrome c_2
2. 2 scorpion neurotoxins
3. chains of a hemerythrin
4. melittin (bee venom), bombinin (skin of amphibians)
5. eledoisin (venom of cephalopods); physalaemin, phyllomedusin, ranatensin, alytesin, bombesin (skin of amphibians); P substance of mammalian hypothalamus
6. two families of snake neurotoxins; secretin; glucagon
7. amphibian kinins—mammalian kinins
8. coerulein, phyllocoerulein (skin of amphibians); pepsin, gastrin, rennin, cholecystokinin, pancreozymin
9. trypsin, chymotrypsin, elastase; thrombin, plasmin, kallicreins, haptoglobin β, endopeptidase liberating carboxypeptidase A
10. ribonuclease, lysozyme, lactose synthetase
11. prolactins (adenohypophysis), growth hormone, lactogen (placenta)
12. glumitocin, isotocin, mesotocin, oxytocin, arginine–vasopressin, lysine–vasopressin
13. iridines I A, I B and II of trout; salmines A I and A II of salmon; clupeines Y I, Y II and Z of herring
14. pepsitensinogen, angiotensinogen
15. insulin A, insulin B
16. trypsin-inhibitor, kallicrein inactivator
17. ACTH, MSH, LPH (all of C regions, all of V regions, V and C having a remote common ancestor)
18. myoglobins, hemoglobins (including bloodworm hemoglobin)
19. β-chains of TSH and of LH
20. two parts of an immunoglobulin light chain

of their configuration, is of great interest to the evolutionist despite its as yet limited scope. The relationship between a snake neurotoxin and vertebrate glucagon, or between the melittin of bee venom and the bombinin of amphibian skin cannot fail to open new vistas and offer hope for the acquisition of a large body of knowledge concerning primary sequences which shed light on the genealogy of protein lineages. As commutation introduces different signifieds among polypeptide chains of common lineage, enhanced knowledge will reveal a more extended extraction of the biosyntagm from the biosystem, one of the main features of biochemical evolution.

3. Scaffolding proteins

At the level anterior in phylogeny to the subdivision of Metazoa in the annelid superphylum and the echinoderm superphylum, we find, as stated above, several branches. In those branches, the scaffolding is insured mainly by collagen, elastin and chitin–protein structures.

In Porifera (Florkin[73]) the skeleton is composed of crystalline material (spicules, sclerites), of fibres of collagen (spongins) or of both, these components of the skeleton being embedded in a jelly-like substance, the mesoglea, which is situated between the external epithelial layer of cells and the choanocytes lining the internal cavity. Demospongiae have siliceous spicules, or collagen, or both. Keratosa only have collagen fibres. Together with mineral compounds, chitin, collagen and elastin contribute to the scaffolding of coelenterates. The same method of scaffolding persists through one of the superphyla (annelid superphylum, Protostomia) the role of chitin–protein becoming dominant in Arthropoda. Since (generally speaking) the echinoderm superphylum (Deuterostomia) has lost the biosynthesis of chitin (Jeuniaux[74]), the scaffolding function is largely fulfilled, in that branch, by collagen and elastin which become the matrix of the internal skeletons of echinoderms and of chordates.

Save for highly exceptional cases, the skeleton of echinoderms is a calcareous network (71–95 per cent of calcium carbonate, 3–15 per cent of magnesium carbonate) whose pores make up more than half the volume. They are filled with connective tissues (Ubaghs[75]).

In Deuterostomia, the functions of connective tissues, though mainly structural or mechanical, may vary with composition. Connective tissues are made of collagen fibres, elastin fibres, mucopolysaccharides and mineral salts. In the adaptations occurring in the course of animal phylogeny, one or another of these constituents may predominate in the complex and determine its functions. This is true of collagen fibres in tendon or dermis, of elastin fibres in ligaments, of mucopolysaccharides in cartilage, and of calcium salts in bone.

The organic matrix of bone, a structure first acquired by fishes and principally composed of collagen, attracts inorganic ions and stabilizes them in the appropriate positions to form seeds of apatite crystals (literature in Neuman and Neuman[76]). Recently the matrix hypothesis has been tentatively replaced by that concerning the direct production of apatite seeds by the cells of the connective tissue.

The collagen of vertebrates is a trimer (one α_2- and two α_1-chains), one-third of the residues being glycine, one-tenth hydroxyproline and one-twelfth proline. The collagen α_2-chains of rat skin, rat tendon, human skin, baboon skin and chicken skin are homologous as are the α_1-collagen chains of rat skin, rat tendon, human, baboon and rabbit skins, bovine skin, chicken skin and chicken bone. The α_1- and α_2-chains are also homologous (see alignment 59 in *Atlas*[10]).

While in invertebrate collagens the glycine content is remarkably constant, the isology is (as expected) less pronounced than it is between the different vertebrate collagens. On the other hand, invertebrate collagens contain more bound carbohydrate than do the vertebrate collagens.

Chitin–protein is not the only structure found in Protostomia (the annelid superphylum). Resilin, a fibrous protein with rubber-like elasticity is found in insects. It contains dityrosine and trityrosine cross-links and is secreted, along with chitin, by the epidermal cells and deposited extracellularly. In Protostomia another structure is found in the tubes built by polychaete annelids. The solid skeleton of these tubes is formed of sulphur-rich scleroproteins, while several glycoproteins are trapped in the meshes of the scleroprotein structure. Another supporting structure of Protostomia, found mainly in Arthropoda, is silk. The silk filament of the *Bombyx mori* cocoon is made up of long double threads of fibroin cemented with sericin. Both are lipoproteins containing small amounts of fats and waxes. In the coagulation during spinning the macromolecules of fibroin (intramolecular hydrogen bonding) are reoriented to intermolecularly beta-pleated sheets, these being grouped in ribbon-like microfibrils (Dobb *et al.*[77]).

There are three segments in fibroins. Segment I consists largely of alternating glycine and alanine, with some serine. Segment II is mainly glycine, alanine, valine and tyrosine. Segment III has many polar and large neutral amino acids (sequences in *Atlas*[10]). The protection provided by silk is outside the organism, in the forms of cocoons, nests, egg cases, etc. Spiders make great use of silk in different episodes of their existence: cocoon surrounding the eggs or several segments of the web, etc. Composition data of limited interest have made room, with the progress of chemical analysis, for much more interesting sequence data (p. D-307 of *Atlas*[10]).

The exoskeletons of vertebrates have been developed on the basis of keratin acquisition, a transformation product of intracellular materials of the epidermis. Keratin of wool is composed of a variety of polypeptide chains of molecular weight from 10 000 to 51 000. One group of these chains is high in sulfur (–S–S– bonds) and lacks α-helix form while the other groups are lower in sulphur, partly in α-helix form and associated in microfibrils whose helical regions are organized in a ring (see Alignment 61 of *Atlas*[10]).

The acquisition of the properties of scaffolding proteins that organisms take advantage of is an expression of the amino acid content and sequence of those proteins. For instance, the sequence (–Gly–Ser–Gly–Ala–Gly–Ala) induces a β-sheet structure such as that of silk, a material which is used in extramural positions and has to be hard but at the same time flexible. The collagen helix is induced by sequences similar to (–Gly–X–Pro–) and evolves where the material has to be rigid and strong, as in fish scales.

Supporting intracellular and extracellular structures are better defined in terms of amino acid sequences. At the time of this writing, knowledge of these sequences in collagen, in keratin and in silk (see Atlas[10]) points to the conclusion that those sequences, though being homologous in each category are not homologous, i.e. are of different ancestry. This conclusion is consistent with the very different qualities of these supporting materials and with the fact that the sequence of amino acids is in each case the background of the biosemiotic value of the syntagm. The reader will find a wealth of data on extracellular and supporting structures in Volumes 26 A, B and C of this Treatise.

4. Biosyntagms involving a proteolytic bioseme

(a) The trypsin lineage

Proteolysis is linked to very widespread biosemes which are inserted at a higher level than the molecular one in a number of biological contexts. In the first Metazoa, digestive enzymes were probably the first enzymes involved in the system which is active beyond the domain of cell metabolism and which parallels multicellularity. This evolution of proteins leads to commutations among derivatives of endodermal metabolism, not only in the gut but also at the level of the glands derived from it. In vertebrates proteolysis is also involved in the process of blood clotting through the biosemes of thrombin and plasmin. It has several functions (physiological radiations) in sexual reproduction, i.e. in the penetration of the spermatozoa and it accompanies the various lysosome activities such as those accompanying the metamorphosis of Amphibia. Very specialized functions such as those of thyroid cells involve a proteolysis mechanism in the libera-tion of thyroid hormones from the thyroid colloid, etc. Another evolutionary feature of the proteolytic enzymes of higher animal forms, besides the specificity of their origin in specialized cells, is their secretion in zymogen form (so far unknown in invertebrates), which involves a blocking of the bioseme of the active site prior to the secretion and the acquisition of a signal for the enzyme activating the zymogen into enzyme and fitting the structure of that enzyme which is fixed at that level on the zymogen*. Proteolytic activity occurs as frequently in invertebrates but, so far as the author knows, the sequence information on proteolytic enzymes of invertebrates is limited to the detection of a serine-active site peptide in the crayfish[80] and in a sea anemone trypsin-related protease[81].

(b) Mammalian pancreatic proteolytic enzymes: trypsin, chymotrypsin, elastase

The secretion of the mammalian pancreas contains homologous zymogens of these serine enzymes (trypsinogen, chymotrypsinogen, proelastase). The specificity of the biosemes (active site) of the enzymes liberated in the duodenum from these zymogens differs with respect to their signified. The bonds hydrolyzed by trypsin involve the carboxyl groups of arginine and lysine, while those split by chymotrypsin A and B involve the carboxyl group of aromatic amino acids. The bonds split by chymotrypsin C are usually leucine-containing peptide bonds.

The active site of serine proteases (comprising not only the pancreatic proteases but thrombin as well) is characterized in animals by the sequence Gly–Asp–Ser–Gly in which Ser appears as the dominating bioseme, whereas in the serine proteases of microorganisms we generally find the sequence Thr–Ser–Met–Ala. The biosemes of the active site take part, by donating a proton, in the formation of a sign in which they are coupled with the bond that corresponds with their specificity (see literature in Dickerson and Geis[79]).

The homology of different proteins of the trypsin lineage is detailed in Table IV, from complete sequencing. α-Chymotrypsinogen has been recognized as a roughly spherical molecule

"with many parallel extended chains, that are probably of β-structure and with only two short segments of α-helix at residues 164–170 and at 234–245. The catalytic site is a shallow depression with residues 57, 120 and 195, next to a hydrophobic pocket which gives the enzyme its specificity" (Dickerson and Geis[79]).

---

* A similar mechanism of partial proteolytic digestion is found in a number of cases: proteolytic enzymes (pepsinogen, chymotrypsinogen, trypsinogen, procarboxypeptidase, proelastase, prorennin), fibrinogen and plastinogen, proinsulin, thyroglobulin, angiotensinogen, brady-kininogen (see Ottensen[78]).

TABLE IV

PROTEASES RELATED TO TRYPSIN

"Active site residues are His-65, Asp-119 and Ser-234. The acidic side-chain of Asp-226 is thought to determine the specificity of trypsin (as well as that of thrombin and trypsin-like enzyme) for peptide bonds following the basic amino acids lysine and arginine. Serine is present at this position in chymotrypsin and elastase. Positions 257 and 272 are occupied by glycine in most of the vertebrate enzymes. Bulkier amino acids in these positions in elastase partially block the entrance to the cavity where substrates are bound, thus determining the specificity of elastase toward the smaller amino acids. Activation of all of the vertebrate enzymes involves proteolytic cleavage of the peptide bond preceding the residue at position 16. Bovine trypsinogen has six disulphide bonds, bovine chymotrypsinogens A and B have five, elastase has four and α-lytic protease has three. In the fragmentary sequences, three disulphide bonds have been determined for trypsin-like enzyme and two for protease A. The three bonds common to all the complete sequences link positions 45 to 66, 186 to 202 and 228 to 263, and are shown by solid arrows. The bond linking 153 to 242 is found in the vertebrate sequences while that linking 1 to 139 occurs only in the chymotrypsinogens. The bonds linking 22 to 175 and 144 to 278 are unique to trypsinogens. Some fragments of thrombin which are shown on the data page have been omitted from this alignment. In some areas the correspondence of the bacterial sequences to the vertebrate sequences is very vague and the alignment shown here is only one of several quite different possibilities." (Alignment 23, in *Atlas*[10])

139 △ (disulphide bond to position 139 marked above position 1)

	1	2	3	4	5	6	7	8	9	10	11	12	13	14	15	16	17	18	19	20
Trypsinogen bovine	–	–	–	–	–	–	–	–	–	V	D	D	D	D	K	I	V	G	G	Y
Trypsinogen spiny dogfish	–	–	–	–	–	–	–	–	A	P	D	D	D	D	K	I	V	G	G	Y
Trypsin turkey	–	–	–	–	–	–	–	–	–	–	–	–	–	–	–	I/				
Chymotrypsinogen A bovine	C	G	V	P	A	I	Q	P	V	L	S	G	L	S	R	I	V	N	G	E
Chymotrypsinogen B bovine	C	G	V	P	A	I	Q	P	V	L	S	G	L	A	R	I	V	N	G	E
Chymotrypsinogen A pig	C	G	V	P	A	I	P	P	V	L	S	G	L	S	R	I	V/			
Elastase pig	–	–	–	–	–	–	–	–	–	–	–	–	–	–	–	V	V	G	G	T
Thrombin B chain bovine	–	–	–	–	–	–	–	–	–	–	–	–	–	–	–	I	V	E	G	Q
Trypsin-like enzyme *Streptomyces griseus*	–	–	–	–	–	–	–	–	–	–	–	–	–	–	–	V	V	G	G	T
a-Lytic protease Myxobacter 495	–	–	–	–	–	–	–	–	–	–	–	–	–	A	N	I	V	G	G	–
Protease A *Streptomyces griseus*																				
Common to all																	V		G	
Common to vertebrate sequences																	V		G	

(continued)

TABLE IV (continued)

Position marker **175** points above column 2 (position 176). Column 10 = position 30; column 20 = position 40.

	1	2	3	4	5	6	7	8	9	0	1	2	3	4	5	6	7	8	9	0
Trypsinogen bovine	T	C	G	A	N	T	V	P	Y	Q	V	S	L	N	–	–	–	–	–	S
Trypsinogen spiny dogfish	E	C	P	K	H	A	A	P	W/				/B	–	–	–	–	–	–	V
Trypsin turkey													/B	–	–	–	–	–	–	S
Chymotrypsinogen A bovine	E	A	V	P	G	S	W	P	W	Q	V	S	L	Q	D	K	–	–	–	T
Chymotrypsinogen B bovine	D	A	V	P	G	S	W	P	W	Q	V	S	L	Q	D	S	–	–	–	T
Chymotrypsinogen A pig																				
Elastase pig	E	A	Q	R	N	S	W	P	S	Q	I	S	L	Q	Y	R	S	G	S	S
Thrombin B chain bovine	D	A	E	V	G	L	S	P	W	Q	V	M	L	F	R	K	S	–	–	P
Trypsin-like enzyme *Streptomyces griseus*	R	A	Q	G	E	F	M	P	F	–	V	R	L	S	M	G	–	–	–	–
a-Lytic protease Myxobacter 495	–	–	–	–	–	–	–	–	–	I	E	Y	S	I	N	N	–	–	–	–
Protease A *Streptomyces griseus*																		/I	T	T
Common to all																				
Common to vertebrate sequences								P		Q			L							

Position marker **66** points above column 5 (position 45). Column 10 = position 50; column 20 = position 60.

	1	2	3	4	5	6	7	8	9	0	1	2	3	4	5	6	7	8	9	0
Trypsinogen bovine	G	Y	H	F	C	–	–	G	G	S	L	I	N	–	–	–	S	Q	W	V
Trypsinogen spiny dogfish	G	Y	H	F	C	–	–	(G.	G.	S)	L.	I	H	–	–	–	Z	Z/		
Trypsin turkey	G	Y	H	F	C	–	–	G	Z	S	L/									
Chymotrypsinogen A bovine	G	F	H	F	C	–	–	G	G	S	L	I	N	–	–	–	E	N	W	V
Chymotrypsinogen B bovine	G	F	H	F	C	–	–	G	G	S	L	I	S	–	–	–	E	D	W	V
Chymotrypsinogen A pig			/H	F	C	–	–	G	(G.	S)	L/									
Elastase pig	W	A	H	T	C	–	–	G	G	T	L	I	R	–	–	–	Q	N	W	V
Thrombin B chain bovine	Q	E	L	L	C	–	–	G	A	S	L	I	S	–	–	–	D	R	W	V
Trypsin-like enzyme *Streptomyces griseus*	–	–	–	–	C	–	–	G	G	A	L/									
a-Lytic protease Myxobacter 495	–	A	S	L	C	S	V	G	F	S	V	T	R	G	A	T	K	G	F	–
Protease A *Streptomyces griseus*	G	G	S	R	C	S	L	G	F	N/										
Common to all					C			G												
Common to vertebrate sequences					C			G			L	I							W	V

TABLE IV *(continued)*

Position markers: `*` above column 5; `45` (↑) above column 6; `7` above the first column of the second group; `8` above the last column.

	1	2	3	4	5	6	7	8	9	0	1	2	3	4	5	6	7	8	9	0
Trypsinogen bovine	V	S	A	A	H	C	Y	K	S	G	I	Q	V	R	L	–	–	G	Q	D
Trypsinogen spiny dogfish			/A	(A.	H.	C.	Y.	R.	S.	G)	I	Q	V	R.	L	–	–	G	Z	H
Trypsin turkey			/A	A	H	C	Y	K/												
Chymotrypsinogen A bovine	V	T	A	A	H	C	G	V	T	T	S	D	V	V	V	–	A	G	E	F
Chymotrypsinogen B bovine	V	T	A	A	H	C	G	V	T	T	S	D	V	V	V	–	A	G	E	F
Chymotrypsinogen A pig			/A	A	H	C	G	V	T	T	S	D/								
Elastase pig	M	T	A	A	H	C	V	D	R	E	L	T	F	R	V	V	V	G	E	H
Thrombin B chain bovine	L	T	A	A	H	C	L	L	Y	P	(W,	B,	F,	K,	B,	–	–	P)	T	V
Trypsin-like enzyme *Streptomyces griseus*		/T	A	A	H	C	V	(N,	S,	G,	S,	N,	G/							
a-Lytic protease Myxobacter 495	V	T	A	G	H	C	G	T	V	N	A	T·	A	R	I	–	G	G	A	V
Protease A *Streptomyces griseus*	/L	T	A	G	H	C	T	N	I	S	A	S/								
Common to all			A		H	C														
Common to vertebrate sequences			A	A	H	C														

Position markers: `9` above the last column of the first group; `10` above the last column.

	1	2	3	4	5	6	7	8	9	0	1	2	3	4	5	6	7	8	9	0
Trypsinogen bovine	N	I	N	V	–	–	–	–	–	V	E	G	N	Q	Q	F	I	S	–	A
Trypsinogen spiny dogfish	B	I	S	A	–	–	–	–	–	B	(Z,	G.	B.	T,	Z)	Y	I	D	–	S
Trypsin turkey																				
Chymotrypsinogen A bovine	D	Q	G	S	–	–	–	–	–	S	S	E	K	I	Q	K	L	K	–	I
Chymotrypsinogen B bovine	D	Q	G	L	–	–	–	–	–	E	T	E	D	T	Q	V	L	K	–	I
Chymotrypsinogen A pig																				
Elastase pig	N	L	N	Q	–	–	–	–	–	N	N	G	T	E	Q	Y	V	G	–	V
Thrombin B chain bovine	B	B	L	L/	H	S	R	T	R	Y	E	R	K	V	E	K	I	S	M	L
Trypsin-like enzyme *Streptomyces griseus*																				
a-Lytic protease Myxobacter 495	–	V	G	T	–	–	–	–	–	–	–	–	–	–	–	–	–	–	–	–
Protease A *Streptomyces griseus*																				
Common to all																				
Common to vertebrate sequences																				

(continued)

TABLE IV *(continued)*

	1	2	3	4	5	6	7	8	9	11 0	1	2	3	4	5	6	7	8	* 9	12 0
Trypsinogen bovine	S	K	S	I	V	H	P	S	Y	N	S	N	T	L	N	N	–	–	D	I
Trypsinogen spiny dogfish	S	M	(V,	I)	R.	H	P	B	Y	(G,	S =	B.	Y.	L.	B =	N.	–	–	D)	I
Trypsin turkey		/A	L	T	H	P	B	Y/												
Chymotrypsinogen A bovine	A	K	V	F	K	N	S	K	Y	N	S	L	T	I	N	N	–	–	D	I
Chymotrypsinogen B bovine	G	K	V	F	K	N	P	K	F	S	I	L	T	V	R	N	–	–	D	I
Chymotrypsinogen A pig																				
Elastase pig	Q	K	I	V	V	H	P	Y	W	N	T	D	D	V	A	A	G	Y	D	I
Thrombin B chain bovine	D	K	I	Y	I	H	P	R	Y	N	W	K	E	N	L	D	R	–	D	I
Trypsin-like enzyme *Streptomyces griseus*																				
a-lytic protease Myxobacter 495	–	–	–	–	–	–	–	–	–	F	A	A	–	R	V	F	P	G	N	D R
Protease A *Streptomyces griseus*																				
Common to all																			D	
Common to vertebrate sequences																			D	I

	1	2	3	4	5	6	7	8	9	13 0	1	2	3	4	5	6	7	8	1 △ 9	14 0
Trypsinogen bovine	M	L	I	K	L	K	S	A	A	S	L	N	S	R	V	A	S	I	S	L
Trypsinogen spiny dogfish	M	L	I	K	L	S	K	P	A	A	L	N	R	D	V	N	L	I	S	L
Trypsin turkey																				
Chymotrypsinogen A bovine	T	L	L	K	L	S	T	A	A	S	F	S	Q	T	V	S	A	V	C	L
Chymotrypsinogen B bovine	T	L	L	K	L	A	T	P	A	Q	F	S	E	T	V	S	A	V	C	L
Chymotrypsinogen A pig																					
Elastase pig	A	L	L	R	L	A	Q	S	V	T	L	N	S	Y	V	Q	L	G	V	L
Thrombin B chain bovine	A	L	L	K	L	K	R	P	I	E	L	S	D	Y	I	H	P	V	C	L
Trypsin-like enzyme *Streptomyces griseus*																				
a-Lytic protease Myxobacter 495	A	W	V	S	L	T	S	A	Q	T	L	–	–	–	–	–	–	–	–	L
Protease A *Streptomyces griseus*																					
Common to all					L															L
Common to vertebrate sequences	L				L															L

TABLE IV *(continued)*

Column markers: 278 (↑ at residue 4), 15 (residue 10), 242 (↑ at residue 13), 16 (residue 20).

Enzyme	1	2	3	4	5	6	7	8	9	0	1	2	3	4	5	6	7	8	9	0
Trypsinogen bovine	P	T	S	C	A	–	–	S	A	G	T	Q	C	L	I	S	G	W	G	N
Trypsinogen spiny dogfish / Trypsin turkey	P	T	G	C	A	–	–	Y	A	G	E	M	C	L	I	S	G	W	G	(B.
Chymotrypsinogen A bovine	P	S	A	S	D	D	F	A	A	G	T	T	C	V	T	T	G	W	G	L
Chymostrypsinogen B bovine / Chymotrypsinogen A pig	P	S	A	D	E	D	F	P	A	G	M	L	C	A	T	T	G	W	G	K
Elastase pig	P	R	A	G	T	I	L	A	N	N	S	P	C	Y	I	T	G	W	G	L
Thrombin B chain bovine	P	K/				/L	L	H	A	G	F	K	G	R	V	T	G	W	G	N
Trypsin-like enzyme *Streptomyces griseus*																				
a-Lytic protease Myxobacter 495	P	R	–	–	–	–	–	V	A	N	G	S	S	F	V	T	V	R	G	S
Protease A *Streptomyces griseus*																				
Common to all	P																		G	
Common to vertebrate sequences	P																G	W	G	

Column markers: 17 (residue 10), 22 (↑ at residue 15), 18 (residue 20).

Enzyme	1	2	3	4	5	6	7	8	9	0	1	2	3	4	5	6	7	8	9	0
Trypsinogen bovine	–	T	K	S	S	G	T	S	Y	P	D	V	L	K	C	L	K	A	P	I
Trypsinogen spiny dogfish / Trypsin turkey	–	T.	B, G, M)A	V	S	–	G	B	Z	L	Z	C	L	D	A	P	V			
Chymotrypsinogen A bovine	–	T	R	Y	T	N	A	N	T	P	D	R	L	Q	Q	A	S	L	P	L
Chymotrypsinogen B bovine / Chymotrypsinogen A pig	–	T	K	Y	N	A	L	K	T	P	D	K	L	Q	Q	A	T	L	P	I
Elastase pig	–	T	R	–	T	N	G	Q	L	A	Q	T	L	Q	Q	A	Y	L	P	T
Thrombin B chain bovine	R/	T	T	S	V	A	E	V	Q	P	S	V	L/	Q	V	V	N	L	P	L
Trypsin-like enzyme *Streptomyces griseus*																				
a-Lytic protease Myxobacter 495	–	T	E	A	–	–	–	–	–	–	–	–	–	–	–	–	–	–	–	A
Protease A *Streptomyces griseus*																				
Common to all		T																		
Common to vertebrate sequences		T											L						P	

(continued)

References p. 220

TABLE IV *(continued)*

202 (↑ above column 6)

	1	2	3	4	5	6	7	8	9	0¹⁹	1	2	3	4	5	6	7	8	9	0²⁰
Trypsinogen bovine	L	S	N	S	S	C	K	S	A	–	Y	P	G	–	Q	I	T	S	N	M
Trypsinogen spiny dogfish	(L.	S,	D.	A,	E)	C	K.	G	A	–	Y	P	G	–	(M,	I.	T.	N,	N)	M
Trypsin turkey																				
Chymotrypsinogen A bovine	L	S	N	T	N	C	K	K	–	–	Y	W	G	T	K	I	K	D	A	M
Chymotrypsinogen B bovine	V	S	N	T	D	C	R	K	–	–	Y	W	G	S	R	V	T	D	V	M
Chymotrypsinogen A pig																				
Elastase pig	V	D	Y	A	I	C	S	S	S	S	Y	W	G	S	T	V	K	N	S	M
Thrombin B chain bovine	/V	E	R	P	V	C	K/						/R	I	R	I	T	B	B	M
Trypsin-like enzyme *Streptomyces griseus*				/A	A	C	R	S	A	–	–	Y	G	N	E/					/E
a-Lytic protease Myxobacter 495	V	G	A	A	V	C	R	S	G	R	T	T	G	Y	Q	–	–	–	–	–
Protease A *Streptomyces griseus*																				
Common to all						C														
Common to vertebrate sequences						C					Y									M

186 (↑ above column 2)

	1	2	3	4	5	6	7	8	9	0²¹	1	2	3	4	5	6	7	8	9	0²²
Trypsinogen bovine	F	C	A	G	Y	–	–	L	E	G	G	K	–	–	–	–	–	–	–	–
Trypsinogen spiny dogfish	M	C.	V	G	Y	–	–	M	Z	G	G	K.	–	–	–	–	–	–	–	–
Trypsin turkey																				
Chymotrypsinogen A bovine	I	C	A	G	–	–	–	A	S	G	V	–	–	–	–	–	–	–	–	–
Chymotrypsinogen B bovine	I	C	A	G	–	–	–	A	S	G	V									
Chymotrypsinogen A pig																				
Elastase pig	V	C	A	G	–	–	–	G	N	G	V	R	–	–	–	–	–	–	–	–
Thrombin B chain bovine	F	C	A	G	Y	K	P	G	E	G	K	R	–	–	–	–	–	–	–	–
Trypsin-like enzyme *Streptomyces griseus*	I	C	A	G	Y	P	D	T	G	G	V	–	–	–	–	–	–	–	–	–
a-Lytic protease Myxobacter 495	–	C	G	T	I	T	A	K	N	V	T	A	N	Y	A	E	G	A	V	R
Protease A *Streptomyces griseus*																				
Common to all		C																		
Common to vertebrate sequences		C		G						G										

TABLE IV (continued)

263
↑   23   *   24

	1	2	3	4	5	6	7	8	9	0	1	2	3	4	5	6	7	8	9	0
Trypsinogen bovine	-	-	-	-	-	D	S	C	Q	-	-	G	D	S	G	G	P	V	V	-
Trypsinogen spiny dogfish	-	-	-	-	-	D	S	C	(Z.	-	-	G.	B.	S.	G.	G.	P.	V)	V	-
Trypsin turkey																				
Chymotrypsinogen A bovine	-	-	-	-	-	S	S	C	M	-	-	G	D	S	G	G	P	L	V	-
Chymotrypsinogen B bovine	-	-	-	-	-	S	S	C	M	-	-	G	D	S	G	G	P	L	V	-
Chymotrypsinogen A pig																				
Elastase pig	-	-	-	-	-	S	G	C	Q	-	-	G	D	S	G	G	P	L	H	-
Thrombin B chain bovine	-	-	-	-	G	D	A	C	E	-	-	G	D	S	G	G	P	F	V	M
Trypsin-like enzyme *Streptomyces griseus*	-	-	-	-	-	D.	T	C	Q	-	-	G	D	S	G	G	P	M	F/	
a-Lytic protease myxobacter 495	G	L	T	Q	G	N	A	C	M	G	R	G	D	S	G	G	S	W	I	-
Protease A *Streptomyces griseus*						/N	V	C	A	E	P	G	D	S	G	G	S	L/		
Common to all								C				G	D	S	G	G				
Common to vertebrate sequences								C				G	D	S	G	G	P			

153
△   25   26

	1	2	3	4	5	6	7	8	9	0	1	2	3	4	5	6	7	8	9	0
Trypsinogen bovine	-	C	S	G	K	-	-	-	-	L	Q	G	I	V	S	W	G	S	-	-
Trypsinogen spiny dogfish	-	C/									/Z.	G.	I.	V.	S/					
Trypsin turkey																				
Chymotrypsinogen A bovine	-	C	K	K	N	G	A	W	T	L	V	G	I	V	S	W	G	S	S	-
Chymotrypsinogen B bovine	-	C	Q	K	N	G	A	W	T	L	A	G	I	V	S	W	G	S	S	-
Chymotrypsinogen A pig																				
Elastase pig	-	C	L	V	N	G	Q	Y	A	V	H	G	V	T	S	F	V	S	R	L
Thrombin B chain bovine	K	S	P	Y	N	N	R	W	Y	Q	M	G	I	V	S	W	G	E	-	-
Trypsin-like enzyme *Streptomyces griseus*	-	T	S	A	G	Q	A	Q	G	V	M	S	G	G	N	V	Q	S	N	G
a-Lytic protease Myxobacter 495											/L	G	L	T	S	G	G	G		
Protease A *Streptomyces griseus*																				
Common to all																				
Common to vertebrate sequences												G			S					

(continued)

References p. 220

TABLE IV (continued)

	228↑		27														144△	28		
	1	2	3	4	5	6	7	8	9	0	1	2	3	4	5	6	7	8	9	0
Trypsinogen bovine	-	G	C	A	Q	-	-	K	N	K	P	G	V	Y	T	K	V	C	N	Y
Trypsinogen spiny dogfish				/A.	Z.	-	-	R)	B	H	P	G	V	Y	T	R/				
Trypsin turkey																				
Chymotrypsinogen A bovine	-	T	C	S	T	-	-	-	S	T	P	G	V	Y	A	R	V	T	A	L
Chymotrypsinogen B bovine	-	T	C	S	T	-	-	-	S	T	P	A	V	Y	A	R	V	T	A	L
Chymotrypsinogen A pig																				
Elastase pig	-	G	C	N	V	-	-	T	R	K	P	T	V	F	T	R	V	S	A	Y
Thrombin B chain bovine	-	G	C	D	R	-	-	N	G	K	Y	G	F	Y	T	H	V	F	R	K
Trypsin-like enzyme *Streptomyces griseus*	Y	G	C	A	R	P	G	-	-	Y	P	G	V/							
α-Lytic protease Myxobacter 495	N	N	C	G	I	P	A	S	Q	R	S	S	L	F	E	R	L	Q	P	I
Protease A *Streptomyces griseus*	G	N	(C.	S)	R	T	G	T	T	F/										
Common to all			C																	
Common to vertebrate sequences			C															V		

								29				
	1	2	3	4	5	6	7	8	9	0	1	2	3
Trypsinogen bovine	V	S	W	I	K	Q	T	I	A	S	N	-	-
Trypsinogen spiny dogfish				/I	H	Z	T	I	A	S	A	-	-
Trypsin turkey										/S	N	-	-
Chymotrypsinogen A bovine	V	N	W	V	Q	Q	T	L	A	A	N	-	-
Chymotrypsinogen B bovine	M	P	W	V	Q	E	T	L	A	A	N	-	-
Chymotrypsinogen A pig							/I.	A)	L	N	-	-	
Elastase pig	I	S	W	I	N	N	V	I	A	S	N	-	-
Thrombin B chain bovine	L	K	W	I	Q	K	V	I	D	R	L	G	S
Trypsin-like enzyme *Streptomyces griseus*													
a-Lytic protease Myxobacter 495	L	S	Q	Y	G	L	S	L	V	T	G	-	-
Protease A *Streptomyces griseus*													
Common to all													
Common to vertebrate sequences				W									

Amino acid code

A	Ala	G	Gly	N	Asn	V	Val
B	Asx	H	His	P	Rro	M	Trp
C	Cys	I	Ile	Q	Gln	X	?
D	Asp	K	Lys	R	Arg	Y	Tyr
E	Glu	L	Leu	S	Ser	Z	Glx
F	Phe	M	Met	T	Thr	—	Gap

The diachronic epigenetic change of signified among the enzymes considered, making it possible to cut proteins at different places, lies in the diversification of this hydrophobic pocket.

The trypsin molecule contains six disulphide bridges while there are five such bridges in chymotrypsin (four of those at the same location in both proteins). Both chymotrypsin and trypsin are synthesized as inactive precursors or zymogens.

Trypsinogen is converted into trypsin by losing its first residues. The splitting due to enterokinase is insured by the biosemic signification of the Lys–Ile bond which forms a couple with enterokinase and, when trypsin is produced, by trypsin itself. Chymotrypsinogen A is converted to Π-chymotrypsin by the splitting (by trypsin) of the Arg –Ile 16 bond (a bioseme coupled with trypsin's active site). Chymotrypsin itself, when formed, splits the leu 13–ser 14 bond (a bioseme coupled with the active site of chymotrypsin). δ-Chymotrypsin is then formed by the removal of the dipeptide ser 14–arg 15. Another dipeptide Thr 147–Asn 148 is then removed to leave A-chymotrypsin containing 3 peptidic chains: chain A (13 residues), chain B (131 residues) and chain C (98 residues). Chain C contains the active site (res. 193–198). Jukes[82] regards chymotrypsinogen A and trypsinogen as examples of an evolutionary step

"caused by gene duplication followed by a few deletions and many random base changes and marked by conservation of function through the preservation of essential sites."

(c) Trypsinopeptides

Trypsinogen is a pancreatic zymogen. Its activation into trypsin is accompanied by the liberation of an N-terminal peptide in which a bioseme consisting of a sequence of four successive aspartic acids has been found just prior to the strategic Lys–Ile bond which is split during activation. This bioseme has been detected in all species so far studied, except in the primitive lungfish *Protopterus aethiopicus*[83]. This accumulation of negative charges has led Neurath and Dixon[84] to conceive the well known model of the activation of trypsinogen in which the four Asp, interacting with positively charged residues elsewhere along the chain of the enzymatic biosyntagm, or by hydrogen bonding, maintain the particular configuration of the zymogen. The hydrolysis of the Lys–Ile bond and the concomitant liberation of the N-terminal peptide would allow the interaction of two end-biosemes His-46 and Ser-183, the result of which produces the bioseme of the active center of trypsin. Delaage et al.[85] and Abita et al.[86] have studied the kinetic constants of trypsinogen and of model peptides and they have concluded that, instead of being in interaction with the rest of the molecule, the activation peptide must be like a tail floating freely at the surface of the macromolecule. The four Asp represent a protective bioseme, slowing down the hydrolysis and preventing the accidental production of trypsin. Since the latter is a universal activator of all pancreatic zymogens, it is clear that its concentration must be kept at a minimum during intracellular transport, during storage in the zymogen granule and in the course of pancreatic secretion.

The sequences of all known trypsinogen activation peptides are listed in Table V. Peptides from two species of primitive fishes, determined in Neurath's laboratory, have been included for comparison. Although these are phylogenetically remote from the artiodactyls, the four successive Asp are found in the dogfish but not in the lungfish where, nevertheless, an accumulation of three acidic residues is found. This particular sequence undoubtedly plays a role in the maintenance of the particular conformation of the inactive precursor, since it has been found in all species examined so far.

The peptides listed in Table V, at least those belonging to artiodactyls, are produced by the activation of cationic proteins. In fact, an anionic

TABLE V

SEQUENCE OF N-TERMINAL ACTIVATION PEPTIDES

Lungfish *(Protopterus aethiopicus)*[83]	Phe-Pro-Ile-Glu-Glu-Asp-Lys
Dogfish *(Squalus acanthias)*[87]	Ala-Pro-Asp-Asp-Asp-Asp-Lys
Horse *(Equus caballus)*[88]	Ser-Ser-Thr-Asp-Asp-Asp-Asp-Lys
Wild boar *(Sus scrofa)*[89]	Phe-Pro-Thr-Asp-Asp-Asp-Asp-Lys
Elephant seal *(Mirunga leonina)*[90]	Phe-Pro-Thr-Asp-Asp-Asp-Asp-Lys
Pig *(Sus scrofa)*[91]	Phe-Pro-Thr-Asp-Asp-Asp-Asp-Lys
Dromedary *(Camelus dromedarius)*[92]	Val-Pro-Ile-Asp-Asp-Asp-Asp-Lys
Red deer *(Cervus elaphus)*[93]	Phe-Pro-Val-Asp-Asp-Asp-Asp-Lys
	Val-Asp-Asp-Asp-Asp-Lys
Roe deer *(Capreolus capreolus)*[93]	Phe-Pro-Val-Asp-Asp-Asp-Asp-Lys
	Val-Asp-Asp-Asp-Asp-Lys
Sheep *(Ovis aries)*[94]	Phe-Pro-Val-Asp-Asp-Asp-Asp-Lys
	Val-Asp-Asp-Asp-Asp-Lys
Goat *(Capra hircus)*[95]	Phe-Pro-Val-Asp-Asp-Asp-Asp-Lys
	Val-Asp-Asp-Asp-Lys
Ox *(Bos taurus)*[96]	Val-Asp-Asp-Asp-Asp-Lys

trypsinogen has also been isolated from the bovine species[97] and from the pig[98,99]. An anionic trypsinogen is probably present in all species listed but it is generally destroyed by the acid extraction procedure.

The comparison of the peptides listed in Table V shows that the more closely related the species, phylogenetically speaking, the more similar are the peptides. The wild boar and the pig have exactly the same activation peptide, which is not surprising since the pig is derived from the wild boar by domestication not exceeding 2500 years in duration. Four species, the red deer, the roe deer, the sheep and the goat also have exactly the same peptides. Two activation peptides have been encountered in each of these four species, one of them two residues shorter than the other. The possible origin of these two peptides has been discussed in papers from the author's laboratory[93,94]. The hypothesis of a duplication of the trypsinogen gene followed by a deletion of two codons in one of the genes has been invoked to explain the presence of two trypsinogens differing in their N-terminal sequence but otherwise very similar, their separation never having been accomplished by any of the numerous methods used. If these genes were allelic, the possibility exists that some individual would be homozygous for one of the genes. The alternative hypothesis, of one unique gene coding for the longest trypsinogen and the shorter variety appearing by artificial reduction of the protein during the extraction procedure, may be more correct. This question must remain open until further results are obtained.

The variable part of the peptides can be compared from a phylogenetic viewpoint. Species such as dogfish *(Squalus acanthias)* and lungfish *(Protopterus aethiopicus)* are too distant from the artiodactyls to be directly compared, except of course that we note the persistence of the sequence of successive acidic residues which one may expect to find in most of the intermediary species and which certainly are akin to the particular properties of the zymogen. As stated above, animals considered to be closely related show greater resemblances in their peptides. For instance, the two deer, the sheep and the goat have exactly the same peptides. Most of the differences between closely related species can be accounted for by a one-base substitution in the corresponding RNA code-word. Since the species compared are closely related and since their peptides are not very different, one can trace the results of the successive point mutations that have affected the DNA through the amino acid substitutions. A tentative explanation of the course of events is proposed in Fig. 6.

The possible point mutations of the first codon, from horse to deer, sheep

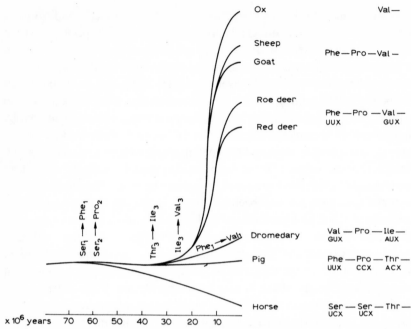

Fig. 6. Classical phylogenic tree of the artiodactyls and the horse and sequence of the variable part of trypsinogen activation peptide. (Bricteux-Grégoire et al.[92])

and goat are: UCX (horse)→UUX (pig)→GUX (dromedary)→UUX (deer, sheep, goat). This implies a retrogressive mutation phenylalanine→valine→phenylalanine. It seems more plausible, however, to assume that phenylalanine has been conserved from the pig to the more recent artiodactyls and that the mutation phenylalanine→valine has occurred only in the Camelidae branch, subsequently to the departure of this branch from the common line. This mutation would thus be of more recent origin.

The second residue, proline, is common to all artiodactyls. A single mutation from UCX coding for serine in the horse accounts for the presence of proline (CCX) in all subsequent species.

The possible point mutations having affected the third codon are: ACX (horse, pig)→AUX (dromedary)→GUX (deer, sheep, goat).

Deletion of the first two codons has occurred in the bovine species.

This is only a tentative and, as such, unsatisfactory explanation of the mutations that may have affected the DNA coding for trypsinogen. One must always be careful in interpreting amino acid substitutions in terms of point mutations, since it is always possible to resort to successive point mutations, to explain the substitution of one residue by another: one has only to postulate intermediaries. However, molecular epigenesis at protein level can only be translated into molecular epigenesis at nucleic acid level if the number of single base substitutions to be invoked as occurring between closely related species is minimal. Before the dromedary sequence was determined the existence of an intermediary between the pig and the deer was postulated, since a single point mutation cannot account for the change of the third residue threonine coded by ACX into valine coded by GUX. This intermediary had to be either isoleucine (coded by AUX) or alanine (coded by GCX). The dromedary case with isoleucine in the third position was thus the "missing link" that confirmed the hypothesis.

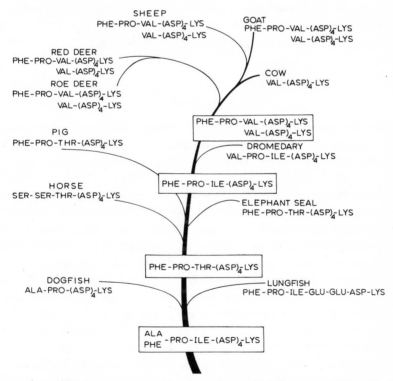

Fig. 7. Evolution of the activation peptides of trypsinogen. The sequences of the hypothetical common ancestors appear in boxes. (Bricteux-Grégoire et al.[90])

The elephant seal sequence (*Mirunga leonina*), belonging to Carnivora was the last to be determined[90]. It is identical to that of porcine trypsinogen.

The pig is considered to be the most primitive among all the artiodactyls studied. It is thus closest to the elephant seal which diverged from the common line some 60 million years ago. It may be assumed that the common ancestor of carnivores and artiodactyls had a trypsinogen beginning with the sequence Phe–Pro–Thr–(Asp)$_4$–Lys. In Fig. 7 the known sequences of trypsinogen activation peptides have been summarized together with classical phylogenetic relationships of the species studied. Ancestral sequences have been assumed by minimizing the number of point mutations required to fit the sequences to the three. This figure has been adapted from the work done by Reeck and Neurath[100] with additional data added. It was assumed that Phe–Pro–Thr was the sequence of the common ancestor of perissodactyls, artiodactyls and carnivores, which mutated later to Phe–Pro–Ile in the artiodactyl branch, before the divergence of the Camelidae, and later on to Phe–Pro–Val before the divergence of Cervidae, Ovidae, and Bovidae. All these modifications can be accounted for by a single point mutation.

(d) Thrombin and plasmin in the context of the evolution of the system of blood clotting in animals

In the hemostatic mechanisms of animals, one common feature is noticeable: it is the aggregation of cells around an injured site, a phenomenon that forms part of the complicated mechanism known as hemostasis.

(i) Hemolymph coagulation in Arthropoda

Duchâteau and Florkin[101] isolated from the hemolymph of *Homarus* a purified, electrophoretically homogeneous "fibrinogen" which is clotted by the simple addition of lobster muscle coagulin, obtained by ultracentrifugation. This coagulin, though partly purified, still consisted, as revealed by electrophoresis, of three components. The instability of the preparation has so far prevented the preparation of the active coagulin factor. Crustacean "fibrinogen" differs from mammalian fibrinogen in solubility and electrophoretic mobility.

In insects, it was observed by Bowen and Kilby[102] that in the locust *Schistocerca gregaria*, of the 770 mg of protein nitrogen per 100 ml of hemolymph plasma, 420 mg were removed by hemolymph clotting.

In the hemolymph of several Crustacea, Hardy[103] discovered a category of fragile blood cells (explosive corpuscles) that are selectively altered on contact with foreign surfaces and which initiate plasma coagulation in their vicinity.

Halliburton[104] interpreted these phenomena by means of the hypothesis that the disruption of the explosive corpuscles releases a factor which determines the coagulation of a coagulable component of the plasma. Tait[105] as well as Tait and Gunn[106] (who confirmed Hardy's observations) agreed with this interpretation.

"Coagulocytes" or fragile hemocytes acting in hemolymph coagulation were identified in *Gryllulus domesticus* and in *Carausius morosus* in the author's laboratory (Grégoire and Florkin[107]) using the phase-contrast microscope and microcinematography. Similar phenomena had been recognized in the hemolymph of *Belostoma fluminea*, with the use of the light microscope by Yeager and Knight[108].

Insect hemolymph coagulation, starting with a cloud of jelly surrounding disrupted coagulocytes (as described by Grégoire and Florkin) was found in a large number of insects by Grégoire who has collected a vast amount of descriptive data on the visual aspects of coagulation in almost 2000 species and proposed a classification of them in several morphological patterns (literature in Grégoire[109]).

However many morphological details may have been described, it is clear that more biochemical data are required before the factors involved in arthropod hemolymph coagulation can be isolated. There is one interesting aspect in the liberation of chemical coagulating factors by the explosive cells of Hardy in Crustacea and coagulocytes in insects in that it shows a feature analogous to the liberation of platelet factors in vertebrate blood clotting.

(ii) Vertebrates

In vertebrates, the coagulation of a specific protein, fibrinogen, is coupled during hemostasis to the cellular reaction that occurs generally in animals and which, in vertebrates, is the function of specialized cells, the platelets, which aggregate and cling to the site of damaged blood vessels. During their aggregation, they release a number of factors.

Fibrinogen is composed of two units, each composed of three peptide chains (A, B, γ). Fibrinogen, a glycoprotein, also contains sialic acid, hexosamine and neutral sugars. The blood also contains a single-chain molecule, prothrombin, which is split up in coagulation into the A and B thrombin chains and possibly in autoprothrombin C (factor X), another proteolytic

TABLE VI

FIBRINOPEPTIDE A

The number of mutations which must have occurred and the number of repeated mutations are shown for each position in this Alignment and in the Alignment of Table VII. This minimum number was calculated using the evolutionary tree of Fig. 8 and the species marked*. (Alignment 20, Atlas[10])

	1	2	3	4	5	6	7	8	9	10	11	12	13	14	15	16	17	18	19	20
Human*	–	–	–	A	D	S	G	–	E	G	D	F	L	A	E	G	G	G	V	R
Chimpanzee	–	–	–	A	D	S	G	–	E	(G.	D)	F	(L.	A.	E.	G.	G.	G.	V)	R
Gorilla	–	–	–	(A.	D.	S.	G.	–	E.	G.	D.	F.	L.	A.	E.	G.	G.	G.	V)	R
Orangutan*	–	–	–	(A.	D.	S.	G.	–	E.	G.	D.	F)	L	(A.	E.	G.	G.	G.	V)	R
Siamang*	–	–	–	A	D	T	G	–	E	G	D	F	L	A	E	G	G	G	V	R
Green monkey*	–	–	–	A	D	T	G	–	E	G	D	F	L	A	E	G	G	G	V	R
Rhesus monkey*	–	–	–	A	D	T	G	–	E	G	D	F	L	A	E	G	G	G	V	R
Irus macaque	–	–	–	A	D	T	G	–	E	G	D	F	L	A	E	G	G	G	V	R
Baboon	–	–	–	A	D	T	G	–	E	G	D	F	L	A	E	G	G	G	V	R
Drill*	–	–	–	(A.	D.	T.	G.	–	D.	G.	D.	F)	I	(T.	E.	G.	G.	G)	V	R
Gibbon	–	–	–	A	D	T	G	–	E	(G.	E)	F	(L.	A.	E.	G.	G.	G.	V)	R
Rabbit*	–	–	–	V	D	P	G	–	E	S	T	F	I	D	E	G	A	T	G	R
Rat*	–	–	–	A	D	T	G	T	T	S	E	F	I	(D,	E.	G.	A,	G,	I)	R
Guinea pig	–	–	–	T	D	T	–	–	–	–	E	F	E	A	A	G	G	G	V	R
Mink	–	–	–	T	N	V	K	–	E	S	E	F	I	A	E	G	A	(A.	G)	R
Badger	–	–	–	T	D	V	K	–	E	S	E	F	I	A	E	G	A	V.	G.	R
Dog*	–	–	–	T	N	S	K	–	E	G	E	F	I	A	E	G	G	G	V	R
Fox	–	–	–	T	N	S	K	–	E	G	E	F	I	A	E	G	G	G	V	R
Brown bear	–	–	–	T	D	G	K	–	E	G	E	F	I	(A.	E.	G.	G.	G.	V)	R
Cat*	–	–	–	G	D	V	Q	–	E	G	E	F	I	A	E	G	G	G	V	R
Lion*	–	–	–	G	D	V	Q	–	E	G	E	F	(I.	A.	E.	G.	G.	G.	V)	R
Gray seal	–	–	–	T	D	T	K	–	E	(S.	D.	F.	L.	A.	E.	G.	G.	G.	V)	R
Horse, mule 1*	–	–	–	–	–	T	E	–	E	G	E	F	L	H	E	G	G	G	V	R
Donkey mule 2*	–	–	–	T	K	T	E	–	E	G	E	F	I	S	E	G	G	G	V	R
Zebra 1*	–	–	–	T	K	T	E	–	E	G	E	F	I	G	E	G	G	(G.	V)	R
Zebra 2	–	–	–	T	K	T	E	–	E	G	E	F	I	S	E	G	(A,	G,	V)	R
Collared peccary	–	–	–	T	T	V	E	E	Q	(S,	Z)	F	(L.	A.	E.	G.	A,	G,	V)	R
Pig, Boar*	–	–	–	A	E	V	Q	D	K	G	E	F	L	A	E	G	G	G	V	R
Llama*	–	T	D	P	D	A	D	–	K	G	E	F	L	A	E	G	G	G	V	R
Vicuna	–	(T.	D.	P.	D.	A.	D.	–	K.	G.	E)	F	(L.	A.	E.	G.	G.	G.	V)	R
Camels*	–	T	D	P	D	A	D	–	E	G	E	F	(L.	A.	E.	G.	G.	G.	V)	R

(continued)

TABLE VI *(continued)*

	1	2	3	4	5	6	7	8	9	0	1	2	3	4	5	6	7	8	9	0
Bovine*	E	D	G	S	D	P	P	-	S	G	D	F	L	T	E	G	G	G	V	R
European bison*	E	D	G	S	D	P	A	-	S	G	D	F	L	A	E	G	G	G	V	R
Cape buffalo*	E	D	G	S	-	-	-	-	-	G	E	F	L	(A.	E.	G.	G.	G.	V)	R
Water buffalo	E	D	G	S	D	A	V	-	G	G	E	F	(L.	A.	E.	G.	G.	G.	V)	R
Persian gazelle*	A	D	D	S	D	(P.	A.	-	G.	G.	E)	F	(L.	A.	E.	G.	G.	G.	V)	R
Sheep*	A	D	D	S	D	P	V	-	G	G	E	F	L	A	E	G	G	G	V	R
Goat	A	D	D	S	D	P	V	-	G	G	E	F	L	A	E	G	G	G	V	R
Ibex	(A.	D.	D.	S.	D.	P.	V.	-	G.	G.	E)	F	(L.	A.	E.	G.	G.	G.	V)	R
Pronghorn*	A	D	G	S	D	P	V	-	G	G	E	S	(L.	P,	D.	G.	A,	T,	G)	R
Reindeer*	A	D	G	S	D	P	A	-	G	G	E	F	L	A	E	G	G	G	V	R
European elk	A	D	G	S	D	P	A	-	G	G	E	F	L	A	E	G	G	G	V	R
Mule deer*	-	-	-	S	D	P	A	-	G	G	E	F	(L.	A.	E.	G.	G.	G.	V)	R
Muntjak*	(A.	D.	G.	S.	D.	P.	A.	-	S.	G.	E)	F	(L.	T.	E.	G.	G.	G.	V)	R
Sika deer	A	D	G	(S.	D.	P.	A.	-	S.	S.	E.	F.	L.	A.	E.	G.	G.	G.	V)	R
Red deer*	A	D	G	S˙	D	P	A	-	S̃	S	D	F	L	A	E	G	G	G	V	R
American elk	(A.	D.	G.	S.	D.	P.	A.	-	S.	S.	D)	F	(L.	A.	E.	G.	G.	G.	V)	R
Kangaroo*	-	-	-	-	-	-	T	K	D	E	G	T	F	I	A	E	G	G	(G.	V) R
Wombat	-	-	-	-	-	-	T	K	T	E	G	S	F	(L.	A.	E.	G.	G.	G.	V) R
Lizard	-	-	-	-	-	-	-	E	D	T	G	T	F	E	E	G	(G.	G,	H,	V) R
Common																G				R
Total mutations	3	3	4	8	6	8	10	3	8	2	5	1	4	8	1	0	2	2	3	0
Repeated mutations	1	1	2	2	2	3	2	1	2	1	3	0	3	2	0	0	1	1	1	0

enzyme which is involved in the final stage of thrombin formation.

Bovine thrombin, a serine proteolytic enzyme (acting like trypsin on arginine and lysine residues) is formed of an A chain (49 residues) linked by a disulphide bond to a B chain (more than 260 residues).

(iii) Fibrinopeptides

Two peptides are released from fibrinogen as a result of the latter's limited proteolysis under the action of thrombin, which splits four Arg–Gly bonds. The fibrinopeptides A and B (negatively charged) are liberated from the terminal regions of A and B chains, respectively. In this operation, arginine–glycine bonds are split and, as a consequence, the terminal amino acid of each of the A and B chains is glycine.

References p. 220

TABLE VII

FIBRINOPEPTIDE B
(Alignment 21, in *Atlas*[10])

	1	2	3	4	5	6	7	8	9	10	11	12	13	14	15	16	17	18	19	20	21	22	Common
Human	-	-	-	-	-	-	-	-	-	Z	G	V	N	D	N	E	E	G	F	F	S	A	R
Chimpanzee	-	-	-	-	-	-	-	-	-	Z	(G.	V.	N.	D.	N.	E.	E.	G.	F)	F	(S.	A)	R
Gorilla	-	-	-	-	-	-	-	-	-	Z	(G.	V.	N.	D.	N.	E.	E.	G.	F.	F.	S.	A)	R
Orangutan	-	-	-	-	-	-	-	-	-	Z	G	V	B	B	B	Z	Z	G	L	(F.	G)	A	R
Siamang	-	-	-	-	-	-	-	-	-	Z	G	V	B	B	B	Z	Z	G	L	(F.	G)	A	R
Gibbon	-	-	-	-	-	-	-	-	-	Z	(G.	V.	B.	B.	B.	–	Z=G.	L)	F	(G.	A)		R
Drill	-	-	-	-	-	-	-	-	-	Z	(G.	V.	B,	G,	B.	E.	E.	G.	L)	F.	G	G	R
Green monkey	-	-	-	-	-	-	-	-	-	Z	(G.	V.	B,	G)	N	E	E	G	L	F	G	G	R
Rhesus monkey	-	-	-	-	-	-	-	-	-	-	-	-	-	-	N	E	E	S	P	F	S	G	R
Irus macaque	-	-	-	-	-	-	-	-	-	-	-	-	-	-	N	E	E	S	P	F	S	G	R
Rabbit	-	-	A	D	D	Y	D	-	-	-	-	-	-	-	D	E	V	L	P	D	A		R
Rat	-	A	T	T	D	S	D	-	-	-	-	-	-	-	K	V	D	(I,	L,	S.	A)		R
Dog	-	-	-	H	Y	Y	D	D	T	D	E	E	E	R	I	V	S	T	V	D	A		R
Fox	-	-	-	(E.	Y.	Y.	D.	D.	T.	D.	E.	E.	E.	R.	I.	V.	S.	T.	V.	D.	A)		R
Cat	-	I	I	D	Y	Y	D	-	E	G	E	E	D	R	D	V	G	V	V	D	A.		R
Lion	-	-	I	D	Y	Y	D	-	(E.	G.	E.	E.	D)	R	(A.	V.	G)	F	(V.	D.	A)		R
Horse, mule 1	-	-	-	L	D	Y	D	H	E	E	E	D	G	R	T	K	V	T	F	D	A		R
Donkey, mule 2	-	-	-	L	D	Y	D	H	E	E	E	D	G	R	T	K	(V.	T.	F.	D.	A)		R
Zebra 1	-	-	-	L	D	Y	D	H	E	E	E	D	G	(R.	A.	K.	V.	T.	F.	D.	A)		R
Zebra 2	-	-	-	L	D	Y	D	H	E	E	E	D	G	(R.	A.	K.	V.	T.	F.	D.	A)		R
Pig	-	-	A	I	D	Y	D	-	E	D	E	D	G	R	P	K	V	H	V	D	A		R
Llama	-	-	A	T	D	Y	D	-	E	E	E	D	D	R	V	(K.	V.	R.	L.	D.	A)		R
Vicuna	-	-	(A.	T.	D.	Y.	D.	-	E.	E.	E.	D.	D)	R	V	K	V	R	(L.	D.	A)		R
Camels	-	-	A	T	D	Y	D	-	E	E	E	D	D	R	V	K	V	R	L	D	A		R
Bovine	Z	F	P	T	D	Y	D	-	E	G	Q	D	D	R	P	K	V	G	L	G	A		R
European bison	(Z.	F.	P.	T.	D.	Y.	D.	-	E.	G.	E.	D.	D.	R.	P.	K)	V	G	L	G	A		R
Cape buffalo	(Z.	F.	P.	T.	D.	Y.	D.	-	E.	G.	E.	D.	D.	R.	P.	K)	S	(G.	L.	G.	A)		R
Water buffalo	Z	(F.	P.	T=D.	Y.	D.	-	E.	G.	Q.	D.	D.	R.	P)	K	(L.	G.	L.	G.	A)			R
Persian gazelle	-	Z	Y	S	D	(Y.	D.	-	E.	V.	E.	E.	D.	R.	P.	K)	S	R	(V.	D.	A)		R
Sheep	-	G	Y	L	D	Y	D	-	E	V	D	D	N	R	A	K	L	P	L	D	A		R
Goat	-	G	Y	L	D	Y	D	-	E	V	D	D	N	R	A	K	L	P	L	D	A		R
Ibex	-	(G.	Y.	L.	D.	Y.	D.	-	E.	V.	D.	D.	N)	R	A	K	(L.	P.	L.	D.	A)		R
Pronghorn	Z	(P,	S=D.	Y.	Y.	D.	-	E.	E.	E.	D.	D)	R	A	K	L	R	(L.	D.	A)			R
Reindeer	Z	H	L	A	D	Y	D	-	E	V	(E.	D.	D)	R	A	K	L	H	L	D	A		R
Mule deer	Z	H	L	(A.	D.	Y.	D.	-	E.	V)	D	D	D	R	A	K	(L.	H.	L)	D	A		R
Muntjak	Z	(H.	S.	T=D.	Y.	D.	-	E.	V.	E.	D.	D)	D	R	A	K	(L.	H.	L.	D.	A)		R
Sika deer	(Z.	H.	S.	T.	D.	Y.	D.	-	E.	E.	E.	E.	D.	R.	A.	K.	L.	H.	L.	D.	A)		R
Red deer	(Z.	H.	S.	T.	D.	Y.	D.	-	E.	E.	E.	E.	D.	R=A	K=L.	H.	L.	D.	A)				R
American elk	Z	(H.	S.	T=D.	Y.	D.	-	E.	E.	E.	E.	D)	R	A	K	(L.	H.	L.	D.	A)			R
Kangaroo	-	S	F	D	Y	D	D	D	T	G	(E,	G,	G,	G,	V,	K,	S,	S,	V,	D,	A)		R
Common																							R
Total mutations	2	8	9	10	4	3	1	3	5	9	6	7	7	3	12	3	10	12	6	5	1	0	
Repeated mutations	1	0	1	3	2	0	0	1	2	5	2	3	3	0	5	1	6	3	3	3	0	0	

Fibrinogen ⟶ 2 A-peptides + 2 B-peptides + fibrin monomer

Fibrin monomers → polymerization → clot

The primary structures of the A and B peptides have been determined in many species (see Doolittle and Blombäck[110]; Blombäck[111]; Dayhoff[10]) and it is clear that in the different species the chains are homologous. We may consider the evolution of the peptides along the phylogenetic tree of artiodactyls. Those species nearest in phylogeny, goat and sheep, present identical sequences of A and B, differing from those of cattle and reindeer, and even more from that of pig. Reindeer is nearer to goat and sheep than these are to cattle. Peptides differ the more as the species involved diverged earlier.

At the COOH end of peptide A, the five species (cattle, sheep, goat, reindeer, pig) have a series of 9 identical residues (9, 11, 12, 14–19). Except in cattle, residue 19 of B is aspartate which may be taken as evidence that glycine was replaced by aspartate after the separation of the cattle branch.

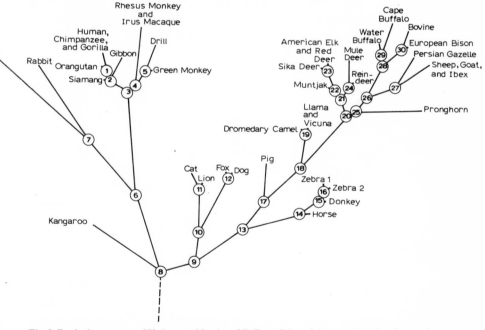

Fig. 8. Evolutionary tree of fibrinopeptides A and B. Branch lengths proportional to the number of amino acid replacements which have occurred, based on Tables VI and VII (Alignments 20 and 21 in the *Atlas*[10]). (*Atlas*[10], Fig. D–6, p. D-89)

In position 18 of fibrinopeptide B we find leucine, except in pig where it is replaced by valine. It thus appears that the substitution of valine by leucine took place after the separation of ruminants and non-ruminants.

Here, as in the case of other proteins, homology is more complete as the species are closer together on the phylogenetic tree. It appears that the fibrinopeptides have evolved rapidly[112]. This is consistent with the limited structural requirements of peptides which need only be strongly acid and must have a terminal arginine residue. The conservative effect of natural selection is limited.

The fibrin of blood clot is digested under the influence of plasmin (fibrinolysin) present in the blood in the form of a zymogen called plasminogen (profibrinolysin) (a single chain). A cleavage at the arginine–valine bond converts it into a molecule composed of two chains (linked by an SH group), a heavy one (± 411 residues) and a light one (± 233 residues). To interpret the function of plasmin as that of making the clot soluble appears to be teleologically naive. It has been claimed that blood contains plasmin inhibitor in sufficient amounts to prevent any plasmin action. But in the clot these inhibitors are lacking and solubilization of the clot still occurs. The fibrinolytic enzyme system has been interpreted recently as a system of inhibition of

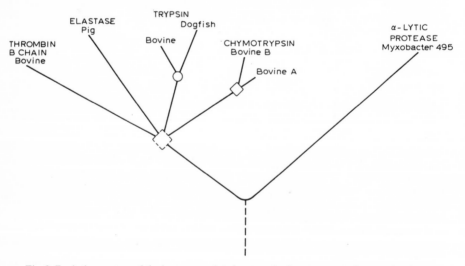

Fig. 9. Evolutionary tree of the proteases related to trypsin. Separate genes for trypsin, chymotrypsin, thrombin and elastase arose from gene duplications which occurred about 1500 million years ago. Genes for chymotrypsins A and B diverged well before the mammalian radiation. (*Atlas*[10], Fig. D-9, p. D-102)

coagulation in the capillary blood circulation. This interpretation is based mainly on the presence of an activator of the fibrinolytic enzyme system in the endothelium of capillaries and on the inhibition of clotting by this system, which in this theory is believed to be active in the maintenance of the potency of capillary circulation. According to Guest[113], besides this prevention of capillary blocking, the fibrinolytic enzyme system under specialized circumstances contributes to the reversal of coagulation in large vessels and tissues, as well as being an anticoagulant agent in the general circulation. Sequence studies have demonstrated (literature, see *Atlas*[10]) that the light chain of plasmin and the B chain of thrombin are homologous with pancreatic proteases (trypsin, chymotrypsin, elastase) and are consequently derived from common ancient genes. Dayhoff[10] has identified in the α-lytic protease of *Myxobacter 495* signs of a relatedness with the same ancient genes. However, the ancestral origin of fibrinogens remains unknown.

(e) Other proteins of the trypsin family

We have noted the homology of trypsin, chymotrypsin, elastase, thrombin and plasmin, which may be thought of as having descended from a common ancestor. As illustrated in Fig. 9 the separate genes for trypsin, chymotrypsin, elastase and thrombin arose about 1500 million years ago. The genes for chymotrypsins A and B diverged before the mammalian radiation and both are therefore likely to be found in any present day mammalian group.

Kallicreins are other carriers of a proteolytic bioseme derived from the same ancestors as the pancreatic proteases, and inserted in a different biological context, *viz.* the liberation from kininogens (glycoproteins found in the α-globulin fraction of mammalian blood serum) of kallidin, a plasma kinin (releasing peripheral blood vessels).

Another compound homologous to the trypsin group is the haptoglobin β-chain. Haptoglobin is a quaternary biosyntagm composed of two α- and two β-chains; its biological activity is to combine with free hemoglobin in blood plasma. Haptoglobin is a glycoprotein with carbohydrate bound to the β-chain. There are reasons for suspecting that haptoglobin is endowed with proteolytic activity involved in the degradation of the hemoglobin with which it combines.

Among the compounds homologous to the trypsin group there is an enzyme involved in the activation of procarboxypeptidase A, a zymogen which is a complex of one molecule of fraction I (precursor of carboxypeptidase A) and two molecules of fraction II (precursor of an endopeptidase homologous to trypsin). Trypsin cleaves fraction II in the endopeptidase appended to itself and the combined action of trypsin and the endopeptidase convert fraction I into carboxypeptidase A.

(f) Proteolytic enzymes not related to trypsin (section 8 of Table III)

The proteolytic enzymes pepsin, gastricsin and rennin are not homologous with the trypsin group, but their degree of isology shows that they descended from a common ancestor to which cholecystokinin, pancreozymin, coerulein and phyllocoerulein (amphibian skin) are also related. Carboxypeptidases A and B show no isology either to the trypsin group or to the pepsin group; their ancestry remains unknown.

5. The lineage: Ribonuclease–lysozyme–lactose synthetase

Ribonuclease (mol. wt. 13 700) hydrolyses RNA in the duodenum. RNA is an alternating copolymer of ribose, sugar and phosphate. One of the four kinds of purine or pyrimidine bases is attached to each ribose. Ribonuclease cuts the chain at the level of a P–O bond at the far side of a P connected to the 3′-carbon of a ribose ring bearing a pyrimidine, either cytosine or uracil. Lysozyme, a bactericidal agent of wide distribution, lyses a mucopolysaccharide structure of the bacterial cell wall which is an alternating copolymer of N-acetylglucosamine and of N-acetylmuramic acid (NAG–NAM–NAG–NAM, the chains being cross-linked by short peptides). Lysozyme cuts the polymer on the far side of an oxygen atom linking a NAG to a NAM and attached to the C_4 of NAG. These proteins of approximately the same length but carriers of widely different biosemes and activities are homologous (paralogous) proteins as a comparison of their primary structures seems to indicate. (An alignment of lactalbumin and lysozyme is presented in Table VIII.)

   Conformational homology is also very pronounced between these molecules. Lysozyme is the first enzyme whose fine structure has been worked out by X-ray analysis. This magnificent work, and the relevant literature, has been presented with a remarkable descriptive conciseness by Dickerson and Geis[79] to whose book the reader is referred. The molecule is egg-shaped. It contains less helix conformation than myoglobin. The residues 5–15,

24–34 and 88–96 are α-helices. The first 40 residues form a compact core and the succeeding residues form a β-sheet.

"The chain runs straight for residues 41–48 and then doubles back for 49–54 to run antiparallel with itself and forms several hydrogen bonds like those in the antiparallel pleated sheat. After burying two hydrophilic groups (55 and 56) in the bottom of the crevice, it doubles back once more and forms a much less regular third strand to the sheet, before going off in another direction at 61".

TABLE VIII
LACTALBUMIN AND LYSOZYME

Glu-35 and Asp-53 are at the active site of lysozyme. One or both of these residues is changed in each lactalbumin sequence. Disulphide bonds have been determined for bovine lactalbumin and chicken lysozyme. The minimum number of mutations which must have occurred and the number of repeated mutations are shown for each position. The evolutionary tree of Fig. 12 was used for these computations as well as for the generation of ancestral sequences. (Alignment 28 in *Atlas*[10])

	1	2	3	4	5	6 (128)	7	8	9	10	11	12	13	14	15	16	17	18	19	20
Lactalbumin human	K	Z	F	T	K	C	E	L	S	Z	L	L	K	-	-	B	I	B	G	Y
Lactalbumin guinea pig	K	Z	L	T	K	C	A	L	S	H	Z	L	B	-	-	B	L	A	G	Y
Lactalbumin bovine	E	Q	L	T	K	C	E	V	F	R	E	L	K	-	-	D	L	K	G	Y
Lysozyme human	K	V	F	E	R	C	E	L	A	R	T	L	K	R	L	G	M	D	G	Y
Lysozyme chicken	K	V	F	G	R	C	E	L	A	A	A	M	K	R	H	G	L	D	N	Y
Lysozyme turkey	K.	V	Y	G	R	(C.	E.	L.	A.	A.	A.	M)	K.	R.	L	G	L	B	B	Y
Lysozyme Japanese quail	K	V	Y	G	R	C	E	L	A	A	A	M	K	R	H	G	L	D	K	Y
Lysozyme duck	K	V	Y	S	R	C	E	L	A	A	A	M	K	R	L	G	L	D	N	Y
Common to all						C														Y
Common to lactalbumin		Q		T	K	C						L				D			G	Y
Common to lysozyme	K	V			R	C	E	L	A				K	R		G		D		Y
Total mutations	1	1	4	3	1	0	1	1	2	3	3	1	1	1	2	1	2	2	2	0
Repeated mutations	0	0	2	0	0	0	0	0	0	0	0	0	0	0	0	0	0	0	0	0

Sequences of common ancestors

	1	2	3	4	5	6	7	8	9	10	11	12	13	14	15	16	17	18	19	20
Node 1	K	Q		T	K	C	E	L	S	R	E	L	K	-	-	D	L	D	G	Y
Node 2	K	Q		T	K	C	E	L	S	R	E	L	K	-	-	D	L	D	G	Y
Node 3	K	V			R	C	E	L	A	R		L	K	R	L	G	L	D	G	Y
Node 4	K	V	Y	G	R	C	E	L	A	A	A	M	K	R	H	G	L	D	N	Y
Node 5	K	V	Y	G	R	C	E	L	A	A	A	M	K	R	L	G	L	D	N	Y
Node 6	K	V	Y		R	C	E	L	A	A	A	M	K	R	L	G	L	D	N	Y

TABLE VIII *(continued)*

116 ↓

	1	2	3	4	5	6	7	8	9	3/0	1	2	3	4	5	*6	7	8	9	4/0
Lactalbumin human	G	G	I	A	L	P	Z	I	L	C	T	M	F	H	T	S	G	Y	B	T
Lactalbumin guinea pig	R	B	I	T	L	P	Z	W	L	C	I	I	F	H	I	S	G	Y	B	T
Lactalbumin bovine	G	G	V	S	L	P	E	W	V	C	T	T	F	H	T	S	G	Y	D	T
Lysozyme Human	R	G	I	S	L	A	N	W	M	C	L	A	K	W	E	S	G	Y	N	T
Lysozyme chicken	R	G	Y	S	L	G	N	W	V	C	A	A	K	F	E	S	N	F	N	T
Lysozyme turkey	R	(G.	Y.	S.	L.	G.	N.	W.	V.	C.	A.	A)	K.	F	Z	S	N	F	N	T
Lysozyme Japanese quail	Q	G	Y	S	L	G	(B.	W=V.	C.	A.	A)		K	F	E	(S.	B.	F.	B.	T.
Lysozyme duck	R	G	Y	S	L	G	N	W	V	C	A	A	N	Y	E	S	S	F	N	T
Common to all					L					C						S				T
Common to lactalbumin					L	P	E			C			F	H		S	G	Y	D	T
Common to lysozyme		G		S	L		N	W		C		A			E	S			N	T
Total mutations	2	1	2	2	0	2	1	1	3	0	3	3	2	3	2	0	2	1	1	0
Repeated mutations	0	0	0	0	0	0	0	0	1	0	0	0	0	0	0	0	0	0	0	0

Sequences of common ancestors

	1	2	3	4	5	6	7	8	9	0	1	2	3	4	5	6	7	8	9	0
Node 1	G	G	I	S	L	P	E	W		C	T		F	H	T	S	G	Y	D	T
Node 2	R	G	I	S	L	P	E	W		C			F	H		S	G	Y	D	T
Node 3	R	G	I	S	L		N	W		C		A	K		E	S	G	Y	N	T
Node 4	R	G	Y	S	L	G	N	W	V	C	A	A	K	F	E	S	N	F	N	T
Node 5	R	G	Y	S	L	G	N	W	V	C	A	A	K	F	E	S	N	F	N	T
Node 6	R	G	Y	S	L	G	N	W	V	C	A	A	K		E	S		F	N	T

	1	2	3	4	5	6	7	8	9	5/0	1	*2	3	4	5	6	7	8	9	6/0
Lactalbumin human	Z	(A.	I.	V.	Z.	B.	-	-	B,	Z,	S,	T.	Z,	Y)	G	L	F	Z	I	B
Lactalbumin guinea pig	Z	A	I	V	K	B	-	-	S	B	H	K	E	Y	G	L	F	Z	I	B
Lactalbumin bovine	E	A	I	V	E	N	-	-	N	Q	S	T	D	Y	G	L	F	Q	I	N
Lysozyme human	R	A	T	N	Y	N	A	G	D	R	S	T	D	Y	G	I	F	Q	I	N
Lysozyme chicken	Q	A	T	N	R	N	T	-	D	G	S	T	D	Y	G	I	L	Q	I	N
Lysozyme turkey	H	A	T	N	R	(B.	T.	-	B.	G.	S.	T.	B.	Y.	G.	I.	L.	Z.	I.	B
Lysozyme Japanese quail	Z.	A.	T.	B)	R	(B.	T.	-	B.	G.	S.	T.	B.	Y.	G.	I.	L=Z.		I.	B
Lysozyme duck	Q	A	T	N	R	N	T	-	D	G	S	T	D	Y	G	I	L	E	I	N

(continued)

TABLE VIII *(continued)*

						5							*						6
1	2	3	4	5	6	7	8	9	0	1	2	3	4	5	6	7	8	9	0
Common to all
	A				N								Y	G				I	N
Common to lactalbumin
E	A	I	V		N								Y	G	L	F	Q	I	N
Common to lysozyme
	A	T	N		N			D		S	T	D	Y	G	I			I	N
Total mutations
3	0	1	1	3	0	2	1	2	3	1	1	2	0	0	1	1	1	0	0
Repeated mutations
0	0	0	0	0	0	0	0	0	0	0	0	0	1	0	0	0	0	0	0

Sequences of common ancestors

	1	2	3	4	5	6	7	8	9	0	1	2	3	4	5	6	7	8	9	0
Node 1	E	A	I	V	E	N	–	–	N	Q	S	T		Y	G	L	F	Q	I	N
Node 2	E	A	I	V		N	–	–			S	T		Y	G	L	F	Q	I	N
Node 3		A	T	N		N	–	D			S	T	D	Y	G	I	F	Q	I	N
Node 4	Q	A	T	N	R	N	T	–	D	G	S	T	D	Y	G	I	L	Q	I	N
Node 5	Q	A	T	N	R	N	T	–	D	G	S	T	D	Y	G	I	L	Q	I	N
Node 6	Q	A	T	N	R	N	T	–	D	G	S	T	D	Y	G	I	L	Q	I	N

				81↑					7							95↑			8
1	2	3	4	5	6	7	8	9	0	1	2	3	4	5	6	7	8	9	0
Lactalbumin human
S	K	L	W	C	K	S	S	Z	V	P	Z	S	R	B	I	C	B	I	S
Lactalbumin guinea pig
B	K	B	F	C	E	S	S	T	T	V	Z	S	R	B	I	C	B	I	S
Lactalbumin bovine
N	K	I	W	C	K	N	D	Q	D	P	H	S	S	N	I	C	N	I	S
Lysozyme human
S	R	Y	W	C	N	D	G	K	T	P	G	A	V	N	A	C	H	L	S
Lysozyme chicken
S	R	W	W	C	N	D	G	R	T	P	G	S	R	N	L	C	N	I	P
Lysozyme turkey
.S)	R	(W.	W.	C.	B.	B.	G)	R	(T.	P.	G.	S)	K	(B.	L.	C.	B.	I.	P
Lysozyme Japanese quail
.S)	R	W	W	(C.	B.	B.	G)	R	T	P	G	S	R	(B.	L=	C.	B.	I.	P
Lysozyme duck
S	R	W	W	C	D	N	G	K	T	P	G	S	K	N	A	C	G	I	P
Common to all
				C										N		C			
Common to lactalbumin
	K			C								S		N	I	C	N	I	S
Common to lysozyme
S	R		W	C			G		T	P	G			N		C			
Total mutations
2	1	4	1	0	2	2	2	3	2	1	2	1	4	0	2	0	2	1	1
Repeated mutations
1	0	0	0	0	0	1	0	0	0	0	0	0	1	0	0	0	0	0	0

(continued)

TABLE VIII *(continued)*

				81				7							95			8		
	1	2	3	4	5	6	7	8	9	0	1	2	3	4	5	6	7	8	9	0

Sequences of common ancestors

	1	2	3	4	5	6	7	8	9	0	1	2	3	4	5	6	7	8	9	0
Node 1		K		W	C	K		S	Q		P	Z	S	R	N	I	C	N	I	S
Node 2		K		W	C			S		T	P	Z	S	R	N	I	C	N	I	S
Node 3	S	R		W	C	B	B	G	K	T	P	G	S		N	A	C	N	I	S
Node 4	S	R	W	W	C	B	B	G	R	T	P	G	S	R	N	L	C	N	I	P
Node 5	S	R	W	W	C	B	B	G	R	T	P	G	S		N	L	C	N	I	P
Node 6	S	R	W	W	C	B	B	G	K	T	P	G	S		N	A	C	N	I	P

	65									9				77						10
	1	2	3	4	5	6	7	8	9	0	1	2	3	4	5	6	7	8	9	0

	1	2	3	4	5	6	7	8	9	0	1	2	3	4	5	6	7	8	9	0
Lactalbumin human	C	B	K	F	L	B	B	B	I	T	B	B	I	M	C	A	K	K	I	L
Lactalbumin guinea pig	C	B	K	L	L	B	B	B	L	T	B	B	I	M	C	V	K	K	I	L
Lactalbumin bovine	C	D	K	F	L	N	N	D	L	T	N	N	I	M	C	V	K	K	I	L
Lysozyme human	C	S	A	L	L	Q	D	N	I	A	D	A	V	A	C	A	K	R	V	R
Lysozyme chicken	C	S	A	L	L	S	S	D	I	T	A	S	V	N	C	A	K	K	I	V
Lysozyme turkey	.C.	S.	A.	L.	L.	S.	S.	B.	I.	T.	A.	S.	V.	B.	C.	A)	K.	K.	I	A
Lysozyme Japanese quail	.C.	S.	A.	L.	L=S.	S.	B.	I.	T.	A.	S.	V.	B=C.	A)	K	K	I	V		
Lysozyme duck	C	S	V	L	L	R	S	D	I	T	E	A	V	R	C	A	K	R	I	V
Common to all	C				L										C		K			
Common to lactalbumin	C	D	K		L	N	N	D		T	N	N	I	M	C		K	K	I	L
Common to lysozyme	C	S		L	L				I				V		C	A	K			
Total mutations	0	1	2	1	0	3	2	1	2	1	3	2	1	3	0	2	0	2	1	3
Repeated mutations	0	0	0	0	0	0	0	0	1	0	0	0	0	0	0	0	1	0	1	0

Sequences of common ancestors

	1	2	3	4	5	6	7	8	9	0	1	2	3	4	5	6	7	8	9	0
Node 1	C	D	K	F	L	N	N	D		T	N	N	I	M	C		K	K	I	L
Node 2	C	D	K	L	L	N	N	D		T	N	N	I	M	C		K	K	I	L
Node 3	C	S	A	L	L			D	I	T		A	V		C	A	K		I	
Node 4	C	S	A	L	L	S	S	D	I	T	A	S	V	N	C	A	K	K	I	V
Node 5	C	S	A	L	L	S	S	D	I	T	A	S	V	N	C	A	K	K	I	V
Node 6	C	S	A	L	L		S	D	I	T		A	V		C	A	K		I	V

TABLE VIII *(continued)*

(column markers: **11** above position 10; **30** ↑ above position 16; **12** above position 20)

	1	2	3	4	5	6	7	8	9	0	1	2	3	4	5	6	7	8	9	0
Lactalbumin human	–	D	I	K	G	I	B	Y	W	L	A	H	K	A	L	C	T	Z	K	–
Lactalbumin guinea pig	–	D	I	K	G	I	B	Y	W	L	A	H	L	P	K	C	S	B	K	–
Lactalbumin bovine	–	D	K	V	G	I	N	Y	W	L	A	H	K	A	L	C	S	E	K	–
Lysozyme human	–	D	P	Q	G	I	R	A	W	V	A	W	R	N	R	C	Q	N	R	D
Lysozyme chicken	S	D	G	D	G	M	N	A	W	V	A	W	R	N	R	C	K	G	T	D
Lysozyme turkey	S	G	G	B	G	M	B	A	(W.	V.	A.	W)	R.	N	R.	C	K	(G.	T.	B
Lysozyme Japanese quail	S	D	V	H	G	M	N	A	W	V	A	W	R	N	R	C	K	G	T	D
Lysozyme duck	S	D	G	D	G	M	N	A	W	V	A	W	R	N	R	C	R	G	T	D
Common to all					G				W		A					C				
Common to lactalbumin	D				G	I	N	Y	W	L	A	H				C			K	
Common to lysozyme					G			A	W	V	A	W	R	N	R	C				D
Total mutations	1	1	4	4	0	1	1	1	0	1	0	1	2	2	2	0	4	2	2	1
Repeated mutations	0	0	0	0	0	0	0	0	0	0	0	0	0	0	0	0	0	0	0	0

Sequences of common ancestors

	1	2	3	4	5	6	7	8	9	0	1	2	3	4	5	6	7	8	9	0
Node 1	–	D	I	K	G	I	N	Y	W	L	A	H	K	A	L	C	S	E	K	–
Node 2	–	D	I	K	G	I	N	Y	W	L	A	H				C	S	N	K	–
Node 3	–	D			G	I	N	A	W	V	A	W	R	N	R	C		N		D
Node 4	S	D	G	D	G	M	N	A	W	V	A	W	R	N	R	C	K	G	T	D
Node 5	S	D	G	D	G	M	N	A	W	V	A	W	R	N	R	C	K	G	T	D
Node 6	S	D	G	D	G	M	N	A	W	V	A	W	R	N	R	C		G	T	D

(column markers: **6** ↑ above position 8; **13** above position 9)

	1	2	3	4	5	6	7	8	9	0	1
Lactalbumin human	L	E	E	W	L	–	–	C	E	K	L
Lactalbumin guinea pig	L	Z	Z	W	Y	–	–	C	E	A	Q
Lactalbumin bovine	L	D	Q	W	L	–	–	C	E	K	L
Lysozyme human	V	R	Q	Y	V	Q	G	C	–	G	V
Lysozyme chicken	V	Q	A	W	I	R	G	C	–	R	L
Lysozyme turkey	V	H	(A.	W.	I)	R	(G.	C)	–	R.	L
Lysozyme Japanese quail	V	N	A	W	I	R	G	C	–	R	L
Lysozyme duck	V	S	K	W	I	R	G	C	–	R	L

(continued)

References p. 220

TABLE VIII *(continued)*

	1	2	3	4	5	6	7	8	9	0	1
								6↑			
Common to all								C			
Common to lactalbumin	L			W				C	E		
Common to lysozyme	V						G	C			
Total mutations	1	6	3	1	3	2	1	0	1	3	2
Repeated mutations	0	0	0	0	0	0	0	0	0	0	0

Sequences of common ancestors

	1	2	3	4	5	6	7	8	9	0	1
Node 1	L	E	Q	W	L	–	–	C	E	K	L
Node 2	L	E	Q	W		–	–	C	E		L
Node 3	V		Q	W			G	C	–		L
Node 4	V		A	W	I	R	G	C	–	R	L
Node 5	V		A	W	I	R	G	C	–	R	L
Node 6	V			W	I	R	G	C	–	R	L

In lysozyme, all the polar groups are on the surface and the hydrophobic ones buried in the interior. In the case of the first 40 residues, the helices have hydrophobic groups packed together while the β-sheet is hydrophilic. This appears to be essential for the proper folding as the high homology of this part in lysozymes of different sources attests. The active site appears contained in a crevice running along the waist of the egg-shaped molecule. Most of the hydrophobic groups situated at the outside of the molecule line this crevice, making it a plausible bioseme for the binding of the substrate. Besides the homology shown by the primary structure, ribonuclease shows a striking conformational homology with lysozyme. Instead of 129, it has 124 amino acid residues and instead of cutting a polysaccharide chain, it cuts a polyribonucleotide chain.

It has the same ovoid shape as lysozyme, with a crevice along the waist. The configuration is epitomized by Dickerson[79] as follows, after the results of Kartha, Bello and Harker[114]:

"The first regular features starting from the *N*-terminal end of the ribonuclease chain are two short lengths of α-helix, 2–12 and 26–33, packed reasonably close to one another. As expected, these helices have hydrophobic sides facing the interior of the molecule. At residue 42, ribonuclease embarks upon a β-sheet structure but on a more extensive scale than in lysozyme. The first strand of the sheet is laid down by 42–49, before the third and last length of a helix, 50–58. The chain then doubles back to begin the most striking feature of the molecule, a double stranded V of β-sheet which runs the entire length of the molecule, first up with residues 71–92 and then back down with 94–110, with 80–86 lying alongside the earlier strand 42–49. This V framework essentially defines the shape of the molecule and its crevice active site. The last act of the folding process is to bring the tail, 116–124, across the inside of the crevice and to position the catalytically important His 119 at the active site".

The active site involves His 12, His 119 and Lys 41.

(a) Commutation in the diachronic molecular epigenesis from animal lysozyme to lactose synthetase

The mammary gland of the therian and monotreme mammals is an evolutionary derivative of sebaceous glands of the skin[115]. It is characterized biochemically by its biosynthesis of a number of milk-specific proteins, specific fat and the disaccharide lactose (glucose + galactose). One of the characteristic abilities of the cell of the lactating mammary gland is the action of the enzyme system lactose synthetase. The biosynthetic pathway of lactose in the mammary gland is represented in Fig. 10.

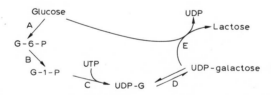

Fig. 10. A, hexokinase; B, phosphoglucomutase; C, UDP-glucose-pyrophosphorylase; D, UDP-galactose-4-epimerase; E, lactose synthetase.

This system is a lateral extension, *via* the acquisition of lactose synthetase, of the biosynthetic pathway which leads *via* UDP–galactose to galacto-lipids[116], to glycoproteins[117,118] and to blood-group substances[119]. In 1966, Brodbeck and Ebner[120] reported on the resolution of a soluble lactose synthetase in two protein components (A and B), neither of which was active *per se* but their combination restoring the activity. The same authors[121] found that the B protein, which is the component with the lower molecular weight, is identical with α-lactalbumin, A + B catalyse the following reaction

$$\text{UDP–Gal + glucose} \rightarrow \text{lactose + UDP} \tag{1}$$

while neither A nor B itself has this property. But, as it was shown by Brew *et al.*[122], A catalyzes the following reaction (as *N*-acetyl-lactosamine synthetase).

$$\text{UDP–D-galactose} + N\text{-acetyl-D-glucosamine} \rightarrow \text{UDP} + N\text{-acetyl-lactosamine} \tag{2}$$

In the presence of increasing amounts of the protein B (α-lactalbumin) reaction (1) is enhanced and reaction (2) inhibited. α-Lactalbumin may therefore be considered as the "specific protein" of lactose synthetase[122].

N-Acetyl-lactosamine is a piece of the glycoprotein-synthesizing machinery (see Brew[123]) and is the subject of a change of signified in the mammary gland *via* the mediation of α-lactalbumin. The primary structure of α-lactalbumin shows, as may be seen in Table VIII, its homology with lysozyme. It therefore appears that α-lactalbumin and lysozyme have arisen by duplication of a common ancestral gene and of the divergent molecular epigenesis of the products.

According to Brew, lysozyme appears as more akin to the ancestral structure and α-lactalbumin as more modified.

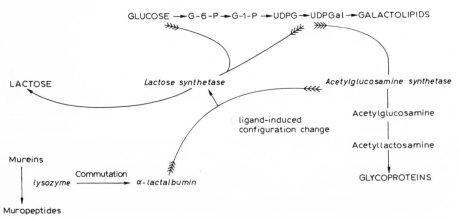

Fig. 11. Origin of a lateral extension of the galactolipid pathway (lactose synthesis). The enzyme *(lactose synthetase)* involved in the new biosyntagm is extracted from the system of protein lineages by a commutation of an enzyme of murein hydrolysis *(lysozyme)* with the production of α-lactalbumin, which produces a ligand-induced configuration change of one of the enzymes of the glycoprotein catenary biosyntagm *(acetylglucosamine synthetase)*, the resulting biocatalyst being *lactose synthetase*. Lactalbumin shows on the other hand, a physiological radiation: its utilization as nutrient for young animals.

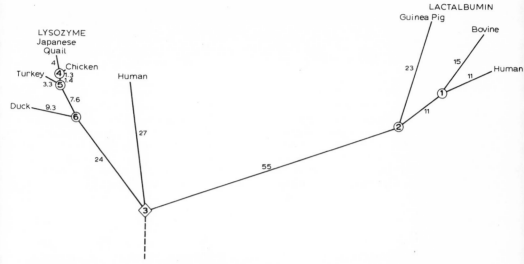

Fig. 12. Evolutionary tree of lactalbumin and lysozyme. Lactalbumin has evolved much faster than lysozyme (rate of change three times higher). Branch lengths in PAMS. (*Atlas*[10], Fig. D-17, p. D-134)

If we compare α-lactalbumin and lysozyme from the point of view of diachronic molecular epigenesis we see that, taking the molecule of lysozyme as the more primitive phylogenetically, residues 64–83, one antigenic site of lysozyme, is the seat of a number of amino acid substitutions in α-lactalbumin, which accounts for the lack of immunological cross-reactivity between the two proteins. In lysozyme, the active region is situated in a cleft in the surface of the protein. This cleft is open at both ends to accommodate a large substrate molecule of oligosaccharide, contact with which is established by amino acid side-chains. In the same region, in α-lactalbumin, a new hydrophobic region cuts out one end of the cleft. This new hydrophobic region appears to result from the substitution of alanine in position 107 of lysozyme by a tyrosine side-chain. The amino acid side-chains which contribute in fixing the large substrate on lysozyme are of no importance in the reaction of α-lactalbumin with its own small substrate. This is true of glutamic acid 35 of lysozyme, replaced by histidine or threonine (depending on how the sequences are aligned). Tryptophan 63 of lysozyme, important for its activity, is also absent in guinea-pig α-lactalbumin. As tryptophan 26 of bovine α-lactalbumin is missing in human α-lactalbumin it cannot be of importance.

The apparently gross difference between the functions of lysozyme and α-lactalbumin may, as suggested by Brew[123] be tentatively considered as a scheme of slow evolution. In this scheme, α-lactalbumin will have, Brew writes

"An activity of transferring an activated galactosyl intermediate from the A protein–UDP-galactose complex to an acceptor characteristic of α-lactalbumin. A type of intermediate which is formed in the lysozyme catalyzed cleavage of an N-acetylneuraminyl-N-acetylglucosaminyl linkage (a C_1 carbonium ion intermediate) could be commonly bound by the active site region of both the A and B proteins. This type of activity is entirely compatible with the modification in the cleft region of α-lactalbumin, and would explain many puzzling features of the lactose synthetase system, including the lack of species specificity in the interaction of the A and B proteins. The apparently very different activities of α-lactalbumin and lysozyme may therefore yet involve basically a very similar reaction mechanism."

The functional epigenesis which results in the lactose synthetase activity appears to reside in a commutation of the specified protein, α-lactalbumin. As underlined by Brew, the evolution of this protein also serves to control lactose synthesis during pregnancy and lactation, by linking together the biosynthesis of milk proteins and carbohydrate.

(b) Rate of epigenesis

The assumption of a constant rate of epigenetic evolution, elaborated in the previous chapter on orthologous protein appears unfounded in the case of paralogous ones. The UEP for lysozyme is in the order of 5.3 M.Y. (Dickerson[112]). If linearity were assumed for α-lactalbumin it would lead to the conclusion that α-lactalbumin diverged from lysozymes 390 M.Y. ago, in the early Devonian, while milk-producing animals emerged 100–150 M.Y. ago. This would mean that commutations introduced a greater amount of new information than can be explained by random point mutations.

6. Insulin [10, 82]

Insulin, the first protein whose primary structure was deciphered, is made of two polypeptide chains, A and B, containing 21 and 30 amino acid residues, respectively. The two chains are covalently linked by two S–S bridges of cysteine residues $(A_7–B_7; A_{20}–B_{19})$. There is another disulphide bridge between A_6 and A_{11}. Chain B shows a very high isology when species are

compared. Chain B in man is identical to the B chain of the elephant, and
differs by only one residue from the B chains of rabbit, ox, pig, dog, sperm-
whale, sheep, goat or horse. It differs from the B chain of the rat by two, and
from the B chain of the guinea pig by 9 changes. Between the B chains of man
and chicken, there are three changes, and 6 residues are different in the B
chain of cod as compared to man.

A chains are identical in man, pig, sperm-whale, dog and rabbit. Differences
between the A chains of man and other mammals concern the four amino
acid residues 4, 8, 9 and 10. A chains of two fishes differ by two residues while
those of man and fish show 9 changes. To compare chains A and B in a given

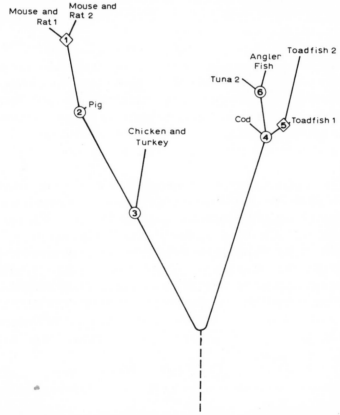

Fig. 13. Evolutionary tree of insulin. Among mammalians, most of the sites do not change, the
observed differences being mainly due to variations at a few very mutable sites. Branch lengths
in PAMs. PAM=the unit of Accepted Point Mutations. One accepted amino acid mutation
for 100 links of protein is one PAM. (*Atlas*[10], Fig. D-22, p. D-176)

species, the chains are aligned and the great degree of isology appears. The two chains may be considered to originate from a single gene. The latter underwent partial duplication and the resulting genes showed base changes (Fig. 13).

7. Hypophyseal hormones[10]

While protein hormones of vertebrates appear to belong to orthologous proteins, the domain of the peptide hormones provides a number of cases of commutation, *i.e.* diachronic molecular epigenesis with a change of signified. The hypophysis was a chemoreceptor before being an endocrine organ (Carlisle[124]). There are three parts in the hypophysis. Two are of ectodermal and one of nervous origin. The anterior part and the intermediate part are derived from Rathke's pouch.

The anterior hypophysis synthesizes a number of hormones.

(*a*) The growth hormone (somatotropin) is a glycoprotein of 188 amino acids, whose primary structure is known in the case of the human hormone. This hormone increases the rate of transport of blood amino acids to tissues and stimulates the incorporation of amino acids into proteins. The antigenic properties are species-specific, which points to interspecific variations.

There is an isological part in somatotropin and in lactogen, a placental hormone.

(*b*) Thyrotropic hormone (thyrotropin, TSH) stimulating the activity of thyroid, is another product of the anterior part of hypophysis. It is also a glycoprotein (15 per cent carbohydrate). Bovine TSH consists of two non-identical polypeptide subunits (Pierce and Liao[125]; Liao and Pierce[126]).

TSH-α has 96 and TSH-β 113 residues. TSH shows a very marked polymorphism. It is situated in two regions of the chain. At the NH_2 terminus most molecules of TSH-α have a Phe–Pro sequence but some terminate with the aspartic acid at position 3. In the case of TSH-β at the COOH terminus some molecules end in Ser–Tyr–Met while others have lost the Met residue. It is possible that some of these polymorphic aspects may be preparation artefacts.

One of the subunits (TSH-α) shows a striking homology (probably identical) with one of the two subunits of LH, LH-α, and to one of the human chorionic gonadotropins.

TSH-β confers the specificity to TSH. Though the amino acid composition of TSH-β and LH-β are quite different, there are 5 regions of isology between them: the dipeptide sequence Lys–Glu (14–15 in TSH-β, 20–21 in LH-β),

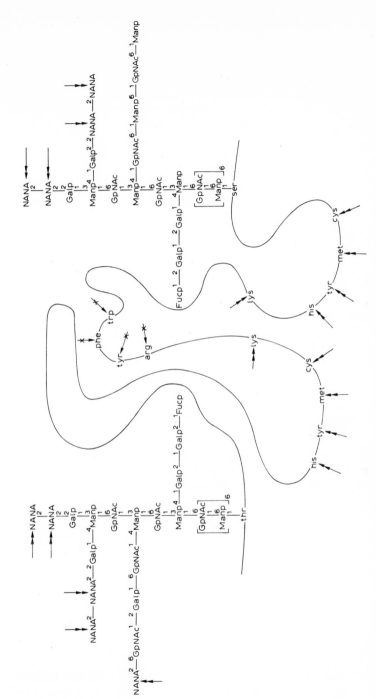

Fig. 14. Schematic representation of the structural data on the human pituitary FSH molecule. →, residues essential for biological activity; →✳, residues not essential for biological activity; NANA, N-acetylneuraminic acid; Galp, galactopyranose, Manp, mannopyranose; Fucp, fucopyranose; GpNAc, 2-acetamido-2-deoxyglucopyranose. The mode of linkage between carbohydrate residues is indicated. *e.g.* Galp 1——4 Manp indicates Galp linked 1→4 to Manp. (Butt and Kennedy[127]).

3 identical tetrapeptide sequences: Ala, Gly, Tyr, Cys (28–31 in TSH-β; 33–36 in LH-2), Val–Cys–Thr–Tyr (51–54 in TSH-β, 55–58 in LH-β); Pro–Gly–Cys–Pro (65–68 in TSH-β, 69–79 in LH-β) and an octapeptide sequence: Val–Ser–Tyr–Pro–Val–Ala–Ileu–Ser (75–82 in TSH-β) and Val–Ser–Phe–Pro–Val–Ala–Leu–Ser (79–86 in LH-β).

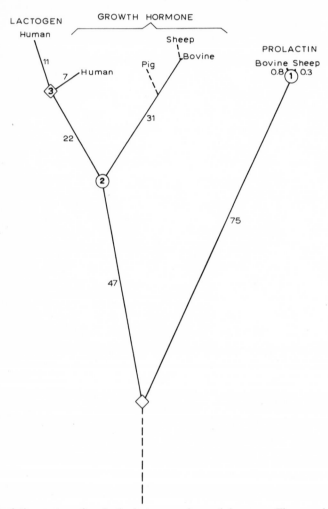

Fig. 15. Evolutionary tree of prolactin, lactogen and growth hormone. The gene duplication which gave rise to the prolactin and growth hormones must have been very ancient. Branch lengths in PAMS. Dashed lines show the approximate location of partial sequences. (*Atlas*[10], Fig. D-20, p, D-175)

(*c*) The third group of anterior hypophysis hormones is formed by gonadotropins of which two are glycoproteins (FSH and LH) and the third is prolactin. FSH (Fig. 14) controls the growth of the follicle in the mammalian ovary, while LH controls its rupture. FSH is also composed of two subunits, non-identical as in LH. One of these, FSH-β, is hormone-specific. FSH-α is highly homologous to LH-α (as well as to HCG-α, human chorionic gonadotropin-α).

(*d*) A fourth principle secreted by the *pars distalis* of the pituitary gland is adrenocorticotropic hormone (ACTH) acting on the adrenal cortex (see Chapter I, Vol. 29A).

(*e*) There is a fifth hormone of the anterior hypophysis, lipotropin (LPH) which stimulates lipolysis. There are two LPH's (α and β). They are highly isologous, as α-LPH consists of the 58 first residues of β-LPH (90 residues). There is also a high degree of isology between LPH's and MSH's, as the latter's sequence is completely present in the LPH's.

8. Neurohypophyseal hormones

The bovine hormones oxytocin and arginine–vasopressin have been known since 1953 thanks to the work done by du Vigneaud[128] and Fromageot[129] and their collaborators. Since then, owing to the work done by Acher and his collaborators[130–134] who have perfected the methods of purification, nine neurohypophyseal hormones listed in Table IX have been isolated. They are all nonapeptides.

Except in the case of cyclostomes, where a single hormone that is chromatographically and pharmacologically similar to arginine–vasotocin seems to be present, two hormones have been identified in vertebrate species. All species except mammals possess arginine–vasotocin and a second hormone. This is either glumitocin, valitocin or aspargtocin as in cartilagenous fishes, isotocin as in bony fishes or mesotocin as in amphibians, reptiles or birds. In mammals, oxytocin and vasopressin are found.

Acher[130] has proposed that isotocin be considered the ancestor of mammal oxytocin, *i.e.* as having appeared in a common ancestor of both, and, by the same token that vasotocin be regarded as an ancestor of arginine–vasopressin and lysine–vasopressin. If isotocin is accepted as the ancestor of oxytocin, the modification of the former to the latter calls for two substitutions only (in 4 and 8). In each category of vertebrates two homologous peptides are found. Their functions may, however, differ as they do, for

TABLE IX

HORMONES OF THE NEUROHYPOPHYSIS[130–134]

Aspargtocin (Squales)

```
1   2   3   4     5       6   7   8       9
Cys-Tyr-Ile-Asn-Asp(NH2)-Cys-Pro-Leu-Gly(NH2)
```

Valitocin (Squales)

```
1   2   3   4     5       6   7   8       9
Cys-Tyr-Ile-Gln-Asp(NH2)-Cys-Pro-Val-Gly(NH2)
```

Glumitocin (Rays)

```
1   2   3   4     5       6   7   8       9
Cys-Tyr-Ile-Ser-Asp(NH2)-Cys-Pro-Gln-Gly(NH2)
```

Isotocin (bony fishes)

```
1   2   3   4     5       6   7   8       9
Cys-Tyr-Ile-Ser-Asp(NH2)-Cys-Pro-Ile-Gly(NH2)
```

Mesotocin (birds, reptiles, amphibians)

```
1   2   3   4     5       6   7   8       9
Cys-Tyr-Ile-Gln-Asp(NH2)-Cys-Pro-Ile-Gly(NH2)
```

Oxytocin (mammals)

```
1   2   3   4     5       6   7   8       9
Cys-Tyr-Ile-Gln-Asp(NH2)-Cys-Pro-Leu-Gly(NH2)
```

Arginine–vasotocin (vertebrates, except mammals)

```
1   2   3   4     5       6   7   8       9
Cys-Tyr-Ile-Gln-Asp(NH2)-Cys-Pro-Arg-Gly(NH2)
```

Arginine–vasopressin (mammals)

```
1   2   3   4     5       6   7   8       9
Cys-Tyr-Phe-Gln-Asp(NH2)-Cys-Pro-Arg-Gly(NH2)
```

Lysine–vasopressin

```
1   2   3   4     5       6   7   8       9
Cys-Tyr-Phe-Gln-Asp(NH2)-Cys-Pro-Lys-Gly(NH2)
```

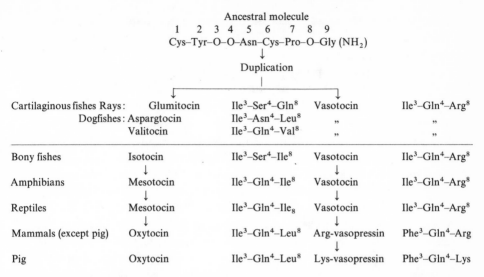

Fig. 16. Hypothetical scheme of the evolution of neurohypophyseal hormones. Following a duplication of the structural gene and a series of point mutations, substitutions in positions 3, 4 and 8 lead to the formation of two evolutionary lineages. (Acher[134])

instance, in vertebrates where oxytocin is active in reproduction and vasopressin in hydromineral regulation.

Acher[134] presents the phylogeny of neurohypophyseal peptides by the scheme of Fig. 16 which suggests the possibility of their derivation from a single ancestor molecule. The two structural genes (if any) controlling the sequence of amino acids in the two peptides may have resulted from the duplication of an ancestral gene.

9. Myoglobin and hemoglobin

(a) Introduction

Primate hemoglobin studies provide an exceptionally valuable collection of data as the comparison of homologous polypeptide chains is here drawn between closely related species, in a relatively short segment of well-established phylogeny. On the other hand, human hemoglobin has been widely studied and the primary structure of its four types of polypeptide chains (α, β, γ, δ) established in the three types of human hemoglobin (HbA $= \alpha\beta$; HbF $= \alpha_2\gamma_2$; HbA$_2 = \alpha_2\delta_2$, see Volume 8 of this Treatise).

Each type of chain is controlled by a separate nonallelic gene. Further-

References p. 220

more, in the study of abnormal hemoglobins point mutations have been described which can be regarded as corresponding to those detected in diachronic epigenesis.

Table X shows the primary structure of α, β and γ chains of human hemoglobins. There are 141 amino acid residues in chain α and 146 in chain β as well as in chain γ. There are human individuals presenting three different categories of variants: *Category 1. Freiburg.* One amino acid is deleted in one chain. Only one case of this is known. Here, valine in position 23 of β-chain is deleted. *Category 2. Lepore.* In this type of abnormal hemoglobin, a hybrid molecule, a normal α-chain is joined to the association of a β-chain and a section of a δ-chain. *Category 3.* The most common type of mutation, which is found at the levels of chain α and chain β consists in the replacement of an amino acid by another at a given point in the chain, a residue carrying an electric charge being replaced by an uncharged residue. This type of mutation, accompanied by an alteration in the electrophoretic migration rate, may be without effect on the properties of the molecule (hemoglobins G, Norfolk or Philadelphia) but the alteration of chain β diminishes the solubility of the hemoglobin and so leads to the grave condition known as sickle-cell anemia.

No advantageous mutation has been identified.

(b) Comparison of hemoglobin chains

In order to determine the degree of similarity and the differences between the polypeptide chains of hemoglobins, we may superimpose their NH_2 ends but must also leave some gaps. At certain points, the β-chain shows the presence of a supplement of amino acid residues. These must be made to correspond with the corresponding gaps in the α chain. As shown in Table X, the polypeptide chains of human hemoglobins are homologous and there is a greater isology between chains γ and β (106 of the 146 residues at same positions) than between chains α and β (58 identical residues). The δ-chain hardly differs (8 of the 146 residues) from chain β.

(c) Tertiary structure

The tertiary structure of hemoglobin depends on a secondary structure of eight α-helical segments (each made of from 6 to 25 amino acids) joined by non-helical segments. The secondary structure depends on the primary

TABLE X

GLOBIN FAMILY—SELECTED SEQUENCES

Positions 52 and 104 are invariant in all known globin sequences. The heme group is coordinated to histidines at positions 73 and 104, except in midge larva globin CTT-III, in which it is coordinated to Ile-77 and His-104. The numbering beneath the alignment applies to sperm whale myoglobin (Alignment 15, *Atlas*[10]).

										1										2
	1	2	3	4	5	6	7	8	9	0	1	2	3	4	5	6	7	8	9	0
α Human	–	–	–	–	–	–	–	–	V	–	L	S	P	A	D	K	T	N	V	K
α Dog	–	–	–	–	–	–	–	–	V	–	L	S	P	A	D	K	T	N	I	K
α Gray kangaroo	–	–	–	–	–	–	–	–	V	–	L	S	A	A	D	K	G	H	V	K
α Chicken	–	–	–	–	–	–	–	–	V	–	L	S	N	A	D	K	N	N	V	K
α Carp	–	–	–	–	–	–	–	–	S	–	L	S	D	K	D	K	A	A	V	K
β Human	–	–	–	–	–	–	–	–	V	H	L	T	P	E	E	K	S	A	V	T
δ Human	–	–	–	–	–	–	–	–	(V.	H.	L.	T.	P.	E.	E)	K.	T	A	V	N
β Dog	–	–	–	–	–	–	–	–	V	H	L	T	A	E	E	K	S	L	V	S
γ Human	–	–	–	–	–	–	–	–	G	H	F	T	E	E	D	K	A	T	I	T
β Gray kangaroo	–	–	–	–	–	–	–	–	V	H	L	T	A	E	E	K	N	A	I	T
β Frog	–	–	–	–	–	–	–	–	–	–	–	–	–	–	G	S	D	L	V	S
Myoglobin sperm whale	–	–	–	–	–	–	–	–	V	–	L	S	E	G	E	W	Q	L	V	L
Myoglobin red kangaroo	–	–	–	–	–	–	–	–	G	–	L	S	D	G	E	W	Q	L	V	L
Globin lamprey	P	I	V	D	S	G	S	V	A	P	L	S	A	A	E	K	T	K	I	R
Globin V sea lamprey	P	I	V	D	T	G	S	V	A	P	L	S	A	A	E	K	T	K	I	R
Globin gastropod mollusc																				
Globin CTT-III midge larva	–	–	–	–	–	–	–	–	–	–	L	S	A	D	Q	I	S	X	V	Z
Leghemoglobin soybean	–	–	–	–	–	–	–	–	V	A	F	T	E	K	E	N	A	L	V	S
									1		2			5					10	
Common to most											L									

										3										4
	1	2	3	4	5	6	7	8	9	0	1	2	3	4	5	6	7	8	9	0
α Human	A	A	W	G	K	V	G	A	H	A	G	E	Y	G	A	E	A	L	E	R
α Dog	S	T	W	D	K	I	G	G	H	A	G	D	Y	G	G	E	A	L	D	R
α Gray kangaroo	A	I	W	G	K	V	G	G	H	A	G	E	Y	A	A	E	G	L	E	R
α Chicken	G	I	F	T	K	I	A	G	H	A	E	E	Y	G	A	E	T	L	E	R
α Carp	I	A	W	A	K	I	S	P	K	A	D	D	I	G	A	E	A	L	G	R
β Human	A	L	W	G	K	V	–	–	N	V	D	E	V	G	G	E	A	L	G	R
δ Human	A	L	W	G	K.	V	–	–	N	V	D	A	V	G	G	E	A	L	G	R
β Dog	G	L	W	G	K	V	–	–	N	V	D	E	V	G	G	E	A	L	G	R
γ Human	S	L	W	G	K	V	–	–	N	V	E	D	A	G	G	E	T	L	G	R
β Gray kangaroo	S	L	W	G	K	V	–	–	A	I	E	Q	T	G	G	E	A	L	G	R
β Frog	G	F	W	G	K	V	–	–	D	A	H	K	I	G	G	E	A	L	A	R

(continued)

TABLE X (continued)

	1	2	3	4	5	6	7	8	9	0 (3)	1	2	3	4	5	6	7	8	9	0 (4)
Myoglobin sperm whale	H	V	W	A	K	V	E	A	D	V	A	G	H	G	Q	D	I	L	I	R
Myoglobin red kangaroo	N	I	W	G	K	V	E	T	D	E	G	G	H	G	K	D	V	L	I	R
Globin lamprey	S	A	W	A	P	V	Y	S	N	Y	E	T	S	G	V	D	I	L	V	K
Globin V sea lamprey	S	A	W	A	P	V	Y	S	N	Y	E	T	S	G	V	D	I	L	V	K
Globin gastropod mollusc																				
Globin CTT-III midge larva	A	S	F	D	K	V	K	-	-	-	-	G	D	P	V	G	I	L	Y	A
Leghemoglobin soybean	S	S	F	E	A	F	K	A	N	I	P	Q	Y	S	X	-	-	Y	T	S
Common to most			W		K								G					L		

(20 marked under column for position 20; 30 marked under column for position 30)

	1	2	3	4	5	6	7	8	9	0 (5)	1	2	3	4	5	6	7	8	9	0 (6)
α Human	M	F	L	S	F	P	T	T	K	T	Y	F	P	H	F	-	D	L	S	H
α Dog	T	F	Q	S	F	P	T	T	K	T	Y	F	P	H	F	-	D	L	S	P
α Gray kangaroo	T	F	H	S	F	P	T	T	K	T	Y	F	P	H	F	-	D	L	S	H
α Chicken	M	F	I	G	F	P	T	T	K	T	Y	F	P	H	F	-	D	L	S	H
α Carp	M	L	T	V	Y	P	Q	T	K	T	Y	F	A	H	W	A	D	L	S	P
β Human	L	L	V	V	Y	P	W	T	Q	R	F	F	E	S	F	G	D	L	S	T
δ Human	(L.	L.	V.	V.	Y.	P.	W.T.		Q)	R	(F.	F.	E.	S.	F.	G.	D.	L.	S.	S.
β Dog	L	L	I	V	Y	P	W	T	Q	R	F	F	D	S	F	G	D	L	S	T
γ Human	L	L	V	V	Y	P	W	T	Q	R	F	F	D	S	F	G	N	L	S	S
β Gray kangaroo	L	L	I	V	Y	P	W	T	S	R	F	F	D	H	F	G	D	L	S	N
β Frog	L	L	V	V	Y	P	W	T	Q	R	Y	F	T	T	F	G	N	L	G	S
Myoglobin sperm whale	L	F	K	S	H	P	E	T	L	E	K	F	D	R	F	K	H	L	K	T
Myoglobin red kangaroo	L	F	K	G	H	P	E	T	L	E	K	F	D	K	F	K	H	L	K	S
Globin lamprey	F	F	T	S	T	P	A	A	Q	E	F	F	P	K	F	K	G	M	T	S
Globin V sea lamprey	F	F	T	S	T	P	A	A	Q.	E)	F	F	P	K	F	K	G	L	T	T
Globin gastropod mollusc																				
Globin CTT-III midge larva	V	F	K	A	D	P	S	I	M	A	K	F	T	Q	F	A	G	-	K	D
Leghemoglobin soybean	(I,	L)	E	K	A	P	A	A	K	D	L	F	S	F	L	A	N	P	T	B
Common to most					P					F			F			L				

(40 marked under column for position 40; 50 marked under column for position 50)

TABLE X (continued)

	1	2	3	4	5	6	7	8	9	7/0	1	2	3	4	5	6	7	8	9	8/0
α Human	-	-	-	-	-	G	S	A	Q	V	K	G	H	G	K	K	V	A	-	D
α Dog	-	-	-	-	-	G	S	A	Q	V	K	A	H	G	K	K	V	A	-	D
α Gray kangaroo	-	-	-	-	-	G	S	A	Q	I	Q	A	H	G	K	K	I	A	-	D
α Chicken	-	-	-	-	-	G	S	A	Q	I	K	G	H	G	K	K	V	A	-	L
α Carp	-	-	-	-	-	G	S	G	P	V	K	-	H	G	K	K	V	I	M	G
β Human	P	D	A	V	M	G	N	P	K	V	K	A	H	G	K	K	V	L	-	G
δ Human	P.	D.	A.	V.	M.	G.	N.	P)	K.	V	K	(A.	H.	G)	K.	K	(V.	L.	-	G
β Dog	P	D	A	V	M	S	N	A	K	V	K	A	H	G	K	K	V	L	-	N
γ Human	A	S	A	I	M	G	N	P	K	V	K	A	H	G	K	K	V	L	-	T
β Gray kangaroo	A	K	A	V	M	A	N	P	K	V	L	A	H	G	A	K	V	L	-	V
β Frog	A	D	A	I	C	H	N	A	K	V	L	A	H	G	E	K	V	L	-	A
Myoglobin sperm whale	E	A	E	M	K	A	S	E	D	L	K	K	H	G	V	T	L	L	-	T
Myoglobin red kangaroo	E	D	E	M	K	A	S	E	D	L	K	K	H	G	I	T	V	L	-	T
Globin lamprey	A	D	E	L	K	K	S	A	D	V	R	W	H	A	E	R	I	I	-	N
Globin V sea lamprey	A	D	Q	L	K	K	S	A	D	V	R	W	H	A	E	R	I	I	-	N
Globin gastropod mollusc																				
Globin CTT-III midge larva	L	E	S	I	K	G	T	A	P	F	H	T	E	A	N	R	I	V	-	G
Leghemoglobin soybean	-	-	-	-	G	V	B	P	K	L	T	G	H	A	E	K	L	F	A	L
Common to most										60			H							70

HEME

	1	2	3	4	5	6	7	8	9/0	1	2	3	4	5	6	7	8	9	10/0	
α Human	A	L	T	N	A	V	A	H	V	D	D	-	M	P	N	A	L	S	A	L
α Dog	A	L	T	T	A	V	A	H	L	D	D	-	L	P	G	A	L	S	A	L
α Gray kangaroo	A	L	G	Q	A	V	E	H	I	D	D	-	L	P	G	T	L	S	K	L
α Chicken	A	I	T	N	A	I	E	H	A	D	D	-	I	S	G	A	L	S	K	L
α Carp	A	V	G	D	A	V	S	K	I	D	D	-	L	V	G	G	L	A	S	L
β Human	A	F	S	D	G	L	A	H	L	D	N	-	L	K	G	T	F	A	T	L
δ Human	A.	F.	S.	D.	G.	L.	A.	H.	L.	D.	N.	-	L)	K.	G	T	F	S	Q	(L.
β Dog	S	F	S	D	G	L	K	N	L	D	N	-	L	K	G	T	F	•A	K	L
γ Human	S	L	G	D	A	I	K	H	L	D	D	-	L	K	G	T	F	A	Q	L
β Gray kangaroo	A	F	G	D	A	I	K	N	L	D	N	-	L	K	G	T	F	A	K	L
β Frog	A	I	G	E	G	L	K	H	P	E	N	-	L	K	A	H	Y	A	K	L

(continued)

References p. 220

TABLE X *(continued)*

HEME									9										10
1	2	3	4	5	6	7	8	9	0	1	2	3	4	5	6	7	8	9	0

	1	2	3	4	5	6	7	8	9	0	1	2	3	4	5	6	7	8	9	0
Myoglobin sperm whale	A	L	G	A	I	L	K	K	K	G	H	–	H	E	A	E	L	K	P	L
Myoglobin red kangaroo	A	L	G	N	I	L	K	K	K	G	H	–	H	E	A	E	L	K	P	L
Globin lamprey	A	V	N	D	A	V	A	S	M	D	D	T	E	K	M	S	M	K	D	L
Globin V sea lamprey	A	V	N	D	A	V	A	S	M	D	D	T	E	K	M	S	M	K	N	L
Globin gastropod mollusc											/K	M	S	A	M	L	S	Q		F
Globin CTT-III midge larva	F	F	S	K	L	I	G	E	I	P	E	N	I	A	D	V	N	T	F	P
Leghemoglobin soybean	V	R	D	S	A	G	Q	L	K	A	S	G	T	V	V	A	D	A	A	L

80

Common to most																				L

									11										12
1	2	3	4	5	6	7	8	9	0	1	2	3	4	5	6	7	8	9	0

	1	2	3	4	5	6	7	8	9	0	1	2	3	4	5	6	7	8	9	0
α Human	S	D	L	H	A	H	K	L	R	V	D	P	V	N	F	K	L	L	S	H
α Dog	S	D	L	H	A	Y	K	L	R	V	D	P	V	N	F	K	L	L	S	H
α Gray kangaroo	S	D	L	H	A	H	K	L	R	V	D	P	V	N	F	K	L	L	S	H
α Chicken	S	D	L	H	A	H	K	L	R	V	D	P	V	N	F	K	L	L	G	Q
α Carp	S	E	L	H	A	S	K	L	R	V	D	P	A	N	F	K	I	L	A	N
β Human	S	E	L	H	C	D	K	L	H	V	D	P	E	N	F	R	L	L	G	N
δ Human	S.	E.	L.	H)	C.	D	K	(L.	H.	V.	D.	P.	E.	N.	F)	R	(L.	L.	G.	N
β Dog	S	E	L	H	C	D	K	L	H	V	D	P	E	N	F	K	L	L	G	N
γ Human	S	E	L	H	C	D	K	L	H	V	D	P	E	N	F	K	L	L	G	N
β Gray kangaroo	S	E	L	H	C	D	K	L	H	V	D	P	E	N	F	K	L	L	G	N
β Frog	S	E	Y	H	S	N	K	L	H	V	D	P	A	N	F	R	L	L	G	N
Myoglobin sperm whale	A	Q	S	H	A	T	K	H	K	I	P	I	K	Y	L	E	F	I	S	E
Myoglobin red kangaroo	A	Q	S	H	A	T	K	H	K	I	P	V	Q	F	L	E	F	I	S	D
Globin lamprey	S	G	K	H	A	K	S	F	Q	V	D	P	Q	Y	F	K	V	L	A	–
Globin V sea lamprey	S	G	K	H	A	K	S	F	Q	V	D	P	Q	Y	F	K	V	L	A	A
Globin gastropod mollusc	A	K	E	H	V	G	–	F	G	V	G	S	A	Q	F	E	N	V	R	S
Globin CTT-III midge larva	A	S	K	H	V	–	–	R	G	V	T	H	B	Z	L	B	B	F	R	A
Leghemoglobin soybean	G	S	V	H	A	Q	–	K	A	V	T	N	P	E	F	–	V	V	K	E

90 100

Common to most				H																

TABLE X *(continued)*

	13										14									
	1	2	3	4	5	6	7	8	9	0	1	2	3	4	5	6	7	8	9	0
α Human	C	L	L	V	T	L	A	A	H	L	P	A	E	F	T	P	A	V	H	A
α Dog	C	L	L	V	T	L	A	C	H	H	P	T	E	F	T	P	A	V	H	A
α Gray kangaroo	C	L	L	V	T	F	A	A	H	L	G	D	A	F	T	P	E	V	H	A
α Chicken	C	F	L	V	V	L	V	A	H	L	P	A	E	L	A	P	K	V	H	A
α Carp	H	I	V	V	G	I	M	F	Y	L	P	G	D	F	P	P	E	V	H	M
β Human	V	L	V	C	V	L	A	H	H	F	G	K	E	F	T	P	P	V	Q	A
δ Human	V.	L.	V)	C	(V.	L.	A)	R.	N	(F.	G)	K	(E.	F.	T.	P.	Q.	M.	Q.	A.
β Dog	V	L	V	C	V	L	A	H	H	F'	G	K	E	F	T	P	Q	V	Q	A
γ Human	V	L	V	T	V	L	A	I	H	F	G	K	E	F	T	P	E	V	Q	A
β Gray kangaroo	I	I	V	I	C	L	A	E	H	F	G	K	E	F	T	I	D	T	Q	V
β Frog	V	F	I	T	V	L	A	R	H	F	Q	H	E	F	T	P	E	L	Q	H
Myoglobin sperm whale	A	I	I	H	V	L	H	S	R	H	P	G	N	F	G	A	D	A	Q	G
Myoglobin red kangaroo	A	I	I	Q	V	I	Q	S	K	H	A	G	N	F	G	A	D	A	Q	A
Globin lamprey	V	I	-	-	-	-	-	-	-	-	-	-	A	D	T	V	A	A	G	D	A
Globin V sea lamprey	V	I	-	-	-	-	-	-	-	-	-	-	A	D	T	V	A	A	G	D	A
Globin gastropod mollusc	M	F	P	G	F	V	A	S	V	A	A	P	P	A	G	A	D	A	-	-
Globin CTT-III midge larva	G	F	V	S	Y	M	K	A	H	-	-	T	D	F	G	A	E	A	-	A
Leghemoglobin soybean	A	L	L	K	T	I	K	A	A	V	G	D	K	W	S	D	E	L	S	R

(position markers: 110, 120)

	15										16													
	1	2	3	4	5	6	7	8	9	0	1	2	3	4	5	6	7	8	9	0	1	2	3	4
Human	S	L	D	K	F	L	A	S	V	S	T	V	L	T	S	K	Y	R	-	-	-	-	-	-
Dog	S	L	D	K	F	F	A	A	V	S	T	V	L	T	S	K	Y	R	-	-	-	-	-	-
Gray kangaroo	S	L	D	K	F	L	A	A	V	S	T	V	L	T	S	K	Y	R	-	-	-	-	-	-
Chicken	S	L	D	K	F	L	C	A	V	G	T	V	L	T	A	K	Y	R	-	-	-	-	-	-
Carp	S	V	D	K	F	F	Q	N	L	A	L	A	L	S	E	K	Y	R	-	-	-	-	-	-
Human	A	Y	Q	K	V	V	A	G	V	A	N	A	L	A	H	K	Y	H	-	-	-	-	-	-
Human	A.	Y.	Q)	K	(V.	V.	A.	G.	V.	A.	N.	A.	L.	A.	H)	K	(Y.	H)	-	-	-	-	-	-
Dog	A	Y	Q	K	V	V	A	G	V	A	N	A	L	A	H	K	Y	H	-	-	-	-	-	-
Human	S	W	Q	K	M	V	T	G	V	A	S	A	L	S	S	R	Y	H	-	-	-	-	-	-
Gray kangaroo	A	W	Q	K	L	V	A	G	V	A	N	A	L	A	H	K	Y	H	-	-	-	-	-	-
Frog	A	L	E	A	H	F	C	A	V	G	D	A	L	A	K	A	Y	H	-	-	-	-	-	-
Myoglobin sperm whale	A	M	N	K	A	L	E	L	F	R	K	D	I	A	A	K	Y	K	E	L	G	Y	Q	G
Myoglobin red kangaroo	A	M	K	K	A	L	E	L	F	R	H	D	M	A	A	K	Y	K	E	F	G	F	Q	G
Globin lamprey	G	F	E	K	L	S	M	(I.	C)	I	L	M	L	R	S	A	Y	-	-	-	-	'	-	-
Globin V sea lamprey	G	F	E	K	L	R	M	I	C	I	L	-	L	R	S	A	Y	-	⌐	-	-	-	-	-
Globin gastropod mollusc	-	W	T	K	L	F	G	L	I	I	D	A	L	K	A	A	G	K	-	-	-	-	-	-
Globin CTT-III midge larva	A	W	G	T	L	(A,	D,	T,	F,	F)	G	M	I	F	S	K	M	-	-	-	-	-	-	-
Leghemoglobin soybean	A	W	E	V	A	Y	D	E	L	A	A	A	I	K	A	K	-	-	-	-	-	-	-	-

(position markers: 130, 140, 150)

| Common to most | | | | K | | | | | | | | | | | Y | | | | | | | | | |

TABLE XI (Hill[135])

Amino acid composition of some tryptic peptides from primate α chains

Human   1 Val-Leu-Ser-Pro-Ala-Asp-Lys-Thr-Asn-Val-Lys.......Met-Phe-Leu-Ser-Phe-Pro-Thr-Thr-Lys...Gly-His-Gly-Lys
(positions 5, 10, 32, 40, 60)

H.lar   (Val, Leu, Ser, Pro, Ala, Asp, Lys) (Thr, Asn, Val, Lys) (Met, Phe, Leu, Ser, Phe, Pro, Thr, Thr, Lys) (Gly, His, Gly, Lys)

P. potto   (Val, Leu, Ser, Pro, Ala, Asp, Lys) (Thr, Asn, Val, Lys)

G. crassicaudatus   (Ser, Asn, Val, Lys)

L. fulvus   (Val, Leu, Ser, Pro, Ala, Asp, Lys) (Thr, Asn, Val, Lys) (Met, Phe, Leu, Ser, Phe, Pro, Thr, Thr, Lys) (Ala, His, Gly, Lys)

L. catta   (Thr, Ans, Val, Lys)

L. variegatus   (Val, Leu, Ser, Pro, Ala, Asp, Lys) (Asp, Asn, Val, Lys) (Met, Phe, Leu, Ser, Phe, Pro, *Leu*, Thr, Lys) (Ala, His, Gly, Lys)

P. verreauxi   (Ser, Asn, *Ileu*, Lys)

Human   62 Val-Ala-Asp-Ala-Leu-Thr-Asn-Ala-Val-Ala-His-Val-Asp-Asp-Met-Pro-Asn-Ala-Leu-Ser-Asp-Leu-His-Ala-His-Lys
(positions 70, 80, 87)

H.lar   (Val, Ala, Asp, Ala, Leu, Thr, Asn, Ala, Val, Ala, His, Val, Asp, Asp, Met, Pro, Asn, Ala, Leu, Ser, Asp, Leu, His, Ala, His, Lys)

G. crassicaudatus   (Val, Ala, Asp, Ala, Leu, Thr, Asn, Ala, Val, Ala, His, Val, Asp, Asp, Met, Pro, Asn, Ala, Leu, Ser, Asp, Leu, His, Ala, His, Lys)

P. potto   (Val, Ala, Asp, Ala, Leu, Thr, Asn, Ala, Val, Ala, His, Val, Asp, Asp, Met, Pro, Asn, Ala, Leu, Ser, Asp, Leu, His, Ala, His, Lys)

L. catta   (Val, Ala, Asp, Ala, Leu, Thr, Asn, Ala, Val, Ala, His, Val, Asp, Asp, Met, Pro, Asn, Ala, Leu, Ser, Asp, Leu, His, Ala, His, Lys)

TABLE XI (continued)

Amino acid composition or sequence of some tryptic peptides from primate β chains

Residues 1–20

γ chains	Gly-His-Phe-Thr[5]-Glu-Gly-Asp-Lys-Ala-Thr-Ileu-Thr-Ser-Leu-Thr-Ser-Leu-Try-Gly-Lys-Val[20]-Asn-Val-Gly-Asp-Ala-Gly-
β chains	Val-His-Leu-Thr-Pro-Gly-Gly-Lys-Ser-Ala-Val-Thr-Ala-Leu-Try-Gly-Lys-Val-Asn-Val-Asp-Glu-Val-Gly-
H.lar	(Val, His, His, Thr, Pro, Glu, Gly, Lys) (Ser, Ala, Val, Thr, Ala, Leu, Try, Gly, Lys) (Val, Asn, Val, Asp, Glu, Val, Gly,
G. crassicaudatus	(Val, His, Phe, Thr, Pro, Gly, Asp, Lys) (Val, Asn, Val, Gly, Glu, Val, Gly,
L. fulvus	Thr-Leu-Leu-Ser-Ala-Glu-Glu-Asp-Ala-His-Val-Thr-Ser-Leu-Try-Gly-Lys-Val-Asn-Val-Asp-Lys-Val-Gly-
L. variegatus	Thr (Leu, Leu, Ser, Ala, Glu, Asp, Asp, Ala, His, Val, Ser, Leu, Try, Gly, Lys) (Val, Asn, Glu, Glu, Val) (Gly,

Residues 25–45

γ chains	Gly-Glu-Thr-Leu-Gly[30]-Arg-Leu-Leu-Val-Val[35]-Tyr-Pro-Try-Thr-Gln-Arg-Phe-Phe-Asp-Ser[45]-Phe-Gly-Asn-Leu-
β chains	Gly-Glu-Ala-Leu-Gly-Arg-Leu-Leu-Val-Val-Tyr-Pro-Try-Thr-Gln-Arg-Phe-Phe-Glu-Ser-Phe-Gly-Asp-Leu-
H.lar	Gly, Glu, Ala, Leu, Gly, Arg) (Leu, Leu, Val, Val, Tyr, Pro, Try, Thr, Gln, Arg) (Phe, Phe, Glu, Ser, Phe, Gly, Asp, Leu,
G. crassicaudatus	Gly, Glu, Ala, Leu, Gly, Arg) (Leu, Leu, Val, Val, Tyr, Pro, Try, Thr, Gln, Arg) (Phe, Phe, Glu, Ser, Phe, Gly, Asp, Leu,
L. fulvus	Gly-Glu-Ala-Leu-Gly-Arg-Leu-Leu-Val-Val-(Tyr, Pro, Try, Thr, Gln, Arg) Phe-Phe-Glu-Glu-Ser(Phe, Glu, Asp, Leu,
L. variegatus	Gly, Glu, Ala, Leu, Gly, Arg) (Leu, Leu, Val, Val, Tyr, Pro, Try, Thr, Gln, Arg) (Phe, Phe, Glu, Ser, Phe, Gly, Asp, Leu,

Residues 50–70

γ chains	Ser-Ser-Ala-Ser-Ala-Ileu-Met-Gly-Asn-Pro-Lys-Val-Lys-Ala-His-Gly-Lys-Lys-Val-Leu-Thr-Ser-Leu-Gly-
β chains	Ser-Thr-Pro-Asp-Ala-Val-Met-Gly-Asn-Pro-Lys-Val-Lys-Ala-His-Gly-Lys-Lys-Val-Leu-Gly-Ala-Phe-Ser-
H.lar	Ser, Thr, Pro, Asp, Ala, Val, Met, Gly, Asn, Pro, Lys) (Val, Lys) (Ala, His, Gly, Lys)
G. crassicaudatus	Ser, Thr, Pro, Glu, Ala, Val, Met, Gly, Asn, Pro, Lys) (Val, Lys) (Ala, His, Gly, Lys, Lys) Lys (Val, Leu, Thr, Ala, Phe, Gly,
L. fulvus	Ser, Ser, Pro, Ala, Ala) (Val, Met, Gly, Asn, Pro, Lys) Val-Lys-Ala-His-Gly-Lys-Lys-Val-Leu-Ser-Val-Leu-Ser (Ala, Phe, Gly,
L. variegatus	Ser, Ser, Pro, Ser, Ala, Ileu, Met, Gly, Asn, Pro, Lys) (Val, Lys) (Ala, His, Gly, Lys, Lys) Lys (Val, Leu, Thr, Ser, Phe, Gly, β

Residues 75–145

γ chains	Asp-Ala-Ileu-Lys-His-Leu-Asp-Asp-Leu-Lys[80]...Met-Val-Thr[135]-Gly-Val-Ala-Ser-Ala-Leu-Ser-Ser-Arg[140]-Tyr-His[145]
β chains	Asp-Gly-Leu-Ala-His-Leu-Asp-Asp-Leu-Lys..Val-Val-Ala-Gly-Val-Ala-Asp-Ala-Leu-Ala-His-Lys-Tyr-His
H.lar	
G. crassicaudatus	Asp, Ala, Val, Ala, His, Leu, Leu, Asp, Asp, Leu, Lys) .. (Val, Val, Ala, Gly, Val, Ala, Asp, Ala, Leu, Ala, His, Lys) (Tyr-His)
L. fulvus	Glu, Ser, Leu, His, His, Leu, Asp, Asp, Leu, Lys) .. Val-Val-Ala-Gly (Val, Ala, Asp, Ala, Leu, Ala, His, Lys) Tyr-His
L. variegatus	Glu, Glu, Thr, Pro, His, Leu, Asp, Asp, Leu, Lys) .. (Val, Val, Thr, Gly, Val, Ala, Asp, Ala, Leu, Ala, His, Lys) (Tyr, His)

structure. Though the primary structures of chains α and β differ, their tertiary structures (as well as that of myoglobin) are very similar. Certain amino acid residues affecting this configuration are found in fixed positions. The proline residues 38 and 102 are at the same place in all chains and this is also true of the hydrophobic residues Leu–Leu in position 107–108 and Val–Leu in 115–116. Certain residues situated near the heme, for instance the phenylalanines 44 and 47, are also invariable, as are the histidines 65 and 94 which react with the heme iron. The remaining regions are more variable.

(d) Molecular diachronic epigenesis in primate hemoglobins

Table XI shows the primary structure of α- and β-chains of the globins of several primates. It appears from the table that the homologous chains of closely related species show a greater isology than those of more distantly related species. In chain β as well as in chain α, long sequences of amino acid residues are common to all primates (for instance, segment 28–42 of β-chain).

The α-chain peptides show few differences. The α-chains of the primitive as well as of the more advanced primates are very isologous to the corresponding peptides of human hemoglobin. It also appears that most of the differences in the β-chains of primates occur at positions where the human β-chain differs from the γ-chain.

When comparing the five human globin polypeptide chains (four in hemoglobin, and myoglobin), we recognize that they have in common twenty-three amino acid positions. These were once considered invariable. But in other species, fourteen other sites appear variable. Thus there are nine sites left as common invariant residues: 26 (Gly), 38 (Pro), 40 (Thr), 44 (Phe), 65 (His), 90 (Leu), 94 (His), 134 (Lys) and 147 (Tyr). Perutz et al.[136] assign a specific function to these invariants. The two histidines (65 and 94) are linked to heme; Leu (90), Phe (44) and Thr (40) are in contact with heme, Thr (40) and Pro (38) are at the corner between the B and C helices; Tyr (147) is internally hydrogen-bonded with the α-CO group of the residue at site 100; Gly (26) is a cross-over point with Gly or Ala at residue 66 of the Σ helix; Lys (134) has no known function.

The invariants account for the tertiary homology of globin chains. Homology, on the other hand, is recognized by a high degree of isology in limited segments of phylogeny and we may postulate that if we were fully informed of the primary structures of all globin chains we would recognize a continuous isology as the basis of the tertiary homology.

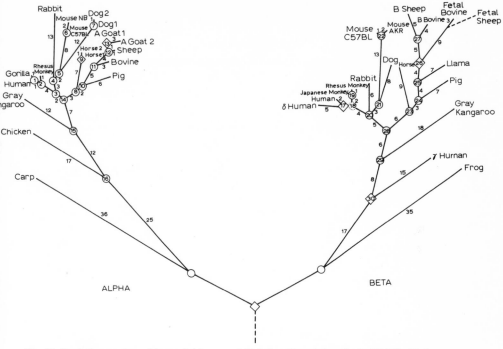

Fig. 17. Evolutionary tree of hemoglobin α- and β-chains. Branch lengths in PAMS. Circles indicate species divergences; squares, gene duplications. (*Atlas*[10], Fig. 3-3, p. 21)

(e) Hemoglobin derivation from myoglobin in diachronic molecular epigenesis

Vertebrate hemoglobin has been considered in section 20 of Chapter I (Vol. 29 A) on the basis of the ligand-induced change of signified it undergoes as a consequence of the fixation of oxygen at the level of one of its heme pockets. The plot of percentage oxygenation *versus* pO_2 which is a rectangular hyperbola in the case of myoglobin becomes a sigmoid curve. The result is that, while myoglobin remains loaded at relatively low oxygen partial pressures, hemoglobin unloads its oxygen at these low pressures.

Myoglobin is homologous with the α-, β-, γ-, and δ-chains of hemoglobins, the synthesis of which is controlled by four different genes. Hemoglobin chains and myoglobin have approximately the same length: 153 amino acid residues in myoglobin; from 141 to 146 in the hemoglobins chains, the shortening being situated at the –COOH end.

The structure and biosemiotic features of myoglobin have been discussed

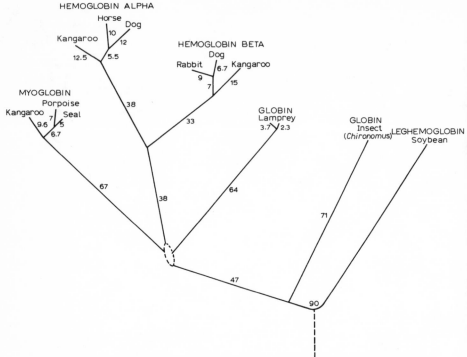

Fig. 18. Evolutionary tree of the globins. Branch lengths are proportional to PAM units. (*Atlas*[10], Fig. 3-2, p. 20)

in section 13 of Chapter I (Vol. 29 A). The isology of myoglobin and hemo-globin chains have led Ingram to postulate that these polypeptides are derived from a common ancestor. A single gene responsible for a primitive hemoglobin-like respiratory pigment may have duplicated, the resulting genes having undergone independent evolution. By a translocation process, the duplicate genes may have been transferred onto different chromosomes. The primitive gene may be likened to the myoglobin-controlling gene. The first non-myoglobin gene may itself have given rise to the three remaining types of hemoglobin genes by dint of duplication and translocation.

Itano[137] and Ingram[138] have advanced a theory concerning the passage from myoglobin to hemoglobin, the cistrons specifying the chains being derived from a common precursor. The situation as we see it today confronts us with five non-allelic cistrons. Ingram[138] has supposed that the hemo-globin molecule was originally rather like the present-day myoglobin (single

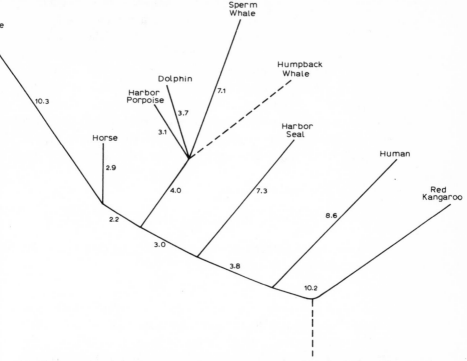

Fig. 19. Evolutionary tree of myoglobin. The branch lengths are in PAM. Humpback whale is represented by a dashed line because only 60 residues of the sequence are known. Aside from the cetacean branches, the sequences represent different mammalian orders; trials with many different topologies clearly showed that this order of divergence gives the minimum tree size. (*Atlas*[10], Fig. 3–6, p. 24)

polypeptide chain, single haem group, no haem–haem interaction, mol. wt. around 17000). He assumes that at that stage (anterior to bony fishes) the haem protein inside muscle cells was the same as in the circulation (α-chain). A gene duplication took place with translocation. One became the modern myoglobin gene, the other the modern α-chain of hemoglobin, which forms, by dimerization, the α_2 molecules, with haem–haem interaction. According to Ingram's speculations the genes of the α_2-chains duplicated again, providing two types of dimers α_2 and γ_2 which in turn form tetramers. The third gene duplication and translocation is believed by Ingram to occur at the level of the γ-chain, the other gene being the modern α-chain of hemoglobin which forms, by dimerization, the α_2 molecules, with haem–haem interaction. Ingram holds that the genes of the α_2-chain duplicated again, providing two

References p. 220

types of dimers α_2 and γ_2, which in turn form tetramers. He says the incidence of natural selection is to be found at the level of the haem interaction which occurs in the dimers and more markedly in the tetramers. The next step in hemoglobin phylogeny, according to Ingram, was the duplication and translocation of the γ-chain gene. The new gene could develop along its own lines providing a tetramer adapted to the physiological needs of the adult organism $(\alpha_2^A \beta_2^A)$ while the old γ-chain continued to develop and provided a hemoglobin adapted to the oxygen exchanges of the foetus $(\alpha_2^A \gamma_2^A)$. At this point three independent genes (α, β, γ) are assumed to have been present, each one capable of forming chains which dimerized and then aggregated to the tetramers $\alpha_2^A \beta_2^A$ or $\alpha_2^A \beta_2^A$.

The succession proposed by Ingram

$$\alpha\text{-chain} \rightarrow \gamma\text{-chain} \rightarrow \beta\text{-chain} \rightarrow \delta\text{-chain}$$

did not stand the test of further sequence studies and the view held today involves the succession

$$\alpha\text{-chain} \rightarrow \beta\text{-chain} \rightarrow \gamma\text{-chain} \rightarrow \delta\text{-chain}$$

The lamprey has a one chain hemoglobin, nearer to myoglobin than any other. The doubling leading to the individualization of α- and β-chain therefore occurred after the lamprey lineage diverged from that of higher vertebrates. This second doubling leading to the α- and β-chains occurred before the appearance of the bony fishes since the carp has an $\alpha_2 \beta_2$ hemoglobin. The first doubling of the gene of the ancestral (myoglobin-like) hemoglobin leading to the individualization of myoglobin (an oxygen container) and of hemoglobin (an oxygen carrier) is very ancient. Even more than by the consideration of amino acid sequences, the homology of myoglobin and hemoglobin is reflected by the fact that each folds its α- and β-chains in identical ways: α is folded as is β in both myoglobin and hemoglobin, (see Vol. 8 of this Treatise).

10. Immunoglobulins

Non-self recognition in vertebrates

The defense of invertebrates against microorganisms is of a non-specific character and is accomplished through phagocytosis and encapsulation as well as by the presence of humoral bactericidal substances. There is no evidence for the existence of an organized lymphoid tissue in invertebrates. Vertebrates have preserved these properties to which they have added the possession of a system of "non-self recognition". DNA informs them of the methods of formation of the self from simple derivatives of the nutrients. This information about the self and the ways of its realization is supplemented with a system of non-self recognition. Constituents foreign to the organism concerned are recognized and eliminated. The foreignness of antigens is recognized by antibodies carrying biosemes for the fixation of non-self compounds. In adult man, for instance, each of the 10^{12} lymphocytes produces one specific antibody. Immunological memory, present in elasmo-branchs and lower bony fishes, is slightly present in lampreys but lacking in hagfish. Bacterial antigens stimulate *agglutinating* antibody production in elasmobranchs, in teleosts, in amphibians and in reptiles but *precipitating* antibodies have rarely been observed. The capacity to form proteins specialized in non-self elimination is a major event of animal biochemical evolution. It evolved in fishes of the Devonian period in parallel with the appearance of organized lymphoid tissue. Protochordates and hagfish possess no such tissues nor any evidence of a lymphoid thymus, whereas this organ is clearly defined in teleost fishes (Salkind[139]).

Vertebrates have acquired—and this is one of their characteristics in biochemical systematics—the ability to respond to antigenic material by the production of specific combining substances (antibodies) as well as the capacity to provide an anamnestic response to these antigens on subsequent exposures. Each lymphocyte is endowed, through hereditary transfer, with a specific gene supply necessary for one species of antigen, or with the repression of a variable part of a common gene. All lymphoid cells derive from the same cells from which all the elements of blood are derived. Certain cells migrate to the thymus and develop as lymphocytes T, while those which mature in the bone marrow become lymphocytes B. The lymphocytes B synthesize the anti-bodies. The presence of an antigen triggers off their multiplication. The lymphocytes T are active, together with the lymphocytes B, in the production of antibodies and they account for cellular immunity.

There are thousands of different (but homologous) immunoglobulins in the serum of an adult vertebrate. In the molecular epigenesis of immuno-globulins, some of the processes involved in the epigenesis of other categories of molecules are recognized. These include gene duplication, duplication within genes, aggregation of several chains (see Chapter I). However, another process, *viz.* the production of a single polypeptidic chain from several genes appears to occur in the case of immunoglobulins.

The synthesis of antibodies, as stated above, takes place in lymphoid cells localized in lymphatic ganglia and in the spleen. During embryonic development, the capacities of a single lymphocyte are reduced to a single immunoglobulin (inhibition of the signified of the other sequencing biosyntagms). The bioseme (minimal significant) of an immunoglobulin is a tetramer formed of 2 heavy chains and 2 light chains. Thousands of heavy and light chains can be produced by the genome of a vertebrate and many combinations are possible. Each cell specializes in one form of heavy and one form of light chain. Some of the combinations are harmful. As a vertebrate organism (sexual reproduction) has two genomes, it is capable of producing antibodies which may destroy the organism. At a certain stage in development, the cells capable of destroying the components of the "self" are eliminated and the system becomes specialized in the recognition and destruction of the "non-self".

When an antigen is introduced in this "non-self" recognizing system, it triggers off the multiplication of cells producing the appropriate antibody. As the antigen has several configurations a heterogeneous collection of antibodies is produced, each specified for a single antigen configuration.

The process becomes more rapid with repetition (chemical memory of the antigen).

In the blood plasma of a vertebrate the immunoglobulins produced by the lymphocytes are recognized as the γ-globulin fraction.

Immunoglobulins are quaternary biosyntagms (heteropolymeric multi-chain proteins) resulting from an assembly of polypeptide chains synthesized in the course of the translation of separate monocistronic mRNAs. The protomer, in all classes of immunoglobulins, is a light-chain–heavy-chain

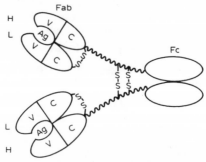

Fig. 20. Schematic model of immunoglobulin molecule. (Grey[140])

pair, containing an antibody-combining site composed of amino acids on both light and heavy chains. In whole molecules there are thus two of these combining sites.

There are disulphide bonds between the two heavy chains and between light and heavy chains. These bonds add to the stability of the molecule but in phylogeny the four-chain structure precedes the existence of the disulphide bridges. The latter are not found, for instance, in *Petromyzon marinus* where the four chains are maintained by non-covalent forces only[142].

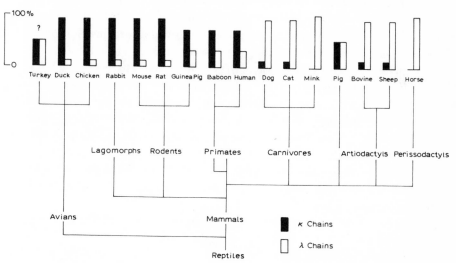

Fig. 21. Distribution of κ and λ light chains among various species. (Hood *et al.*[141])

The different immunoglobulins (mol.wt. 100000–1000000) are aggregates of variable combinations of light and heavy chains. They also contain carbohydrates. There are several types in the γ-globulin, the principal ones being designated as IgA, IgM, IgG, IgD and IgE. The antibodies of the IgM category, of constant and slight proportion, correspond to the natural immunity that stems from the collection of antigens in the environment (food, microorganisms, plants, etc.).

IgG (separable by electrophoresis in several sub-classes) is related to the defense function of the organism. It is a tetramer of two light chains and two heavy chains (50000–55000). The light chains are either of type κ or of type λ.

In the light chains, there are two S–S bridges, while in each heavy γ-chain there are 3 or 4 S–S bridges.

In the five classes of immunoglobulins, each heavy chain is associated with a λ or with a $\varkappa$ light chain.

The light chains of vertebrates are similar in size and have a molecular weight of 20000–25000. By analytical ultracentrifugation, Clemm and Small[143] have determined a value of approximately 23000 for lemon shark and rabbit light chains. Bullfrog *(Rana catesbeiana)* (Marchalonis and Edelman[144]) are of similar molecular weight, as well as those of men and mice (Titani *et al.*[145]; Gray[146]).

Both the light chains and the heavy chains contain a constant part (C) and a variable part (V). It is believed that the constant part of the chains is the result of the presence, in all lymphoid cells, of a special sequencing biosyntagm which is held to be three times longer in heavy chains than in light chains.

With respect to the production of the variable parts (110 residues in heavy as well as in light chains) several interpretations have been suggested. One of them is based on genome heterogeneity and the postulated presence in each lymphoid cell of a large number of V genes resulting from both duplication and independent epigenesis. One gene for a heavy V chain and one gene for a light V chain would be activated and all other ones repressed. Another view has it that mechanisms operating during somatic development would diversify a single V gene. Yet another theory postulates a diversification of the translation process of one and the same V gene.

The progress of sequence data collection has revealed an internal homology in units of 11000–12000 (one half of a light chain[147]) put together to form both heavy and light chains.

"Since the most primitive vertebrates extant have immunoglobulins consisting of fully developed heavy and light chains the origin of the primitive half light chain is placed in the prevertebrate era. Through partial gene duplication, a gene capable of coding for a peptide chain the length of a light chain evolved and, subsequent to this, other gene duplication occurred so that two light chain genes evolved which eventually diverged to form the $\varkappa$ and λ genes. By partial gene duplication, a gene capable of coding for a peptide chain equal to one and a half times the length of a light chain was formed (V_1C_2). This was the primitive heavy chain gene. Subsequent gene duplications led to the formation of peptide chains up to two and a half times the size of a light chain (V_1C_4)." (Grey[148]).

Two types of light chains have been detected (by serologic studies and peptide fingerprinting) in several mammals, and Hood *et al.*[141] have studied the distribution shown in Fig. 21, of $\varkappa$ and λ light chains among various species of vertebrates. In all species (except mink and horse) there is evidence for the two types of light chains, phylogenetically closely related species tending to have similar $\varkappa/\lambda$ ratios. In the leopard shark one-third of the light chains are of the $\varkappa$ type, as shown by Suran and Papermaster[149]. Immuno-globulins are biosyntagms the signified of which results in the association of

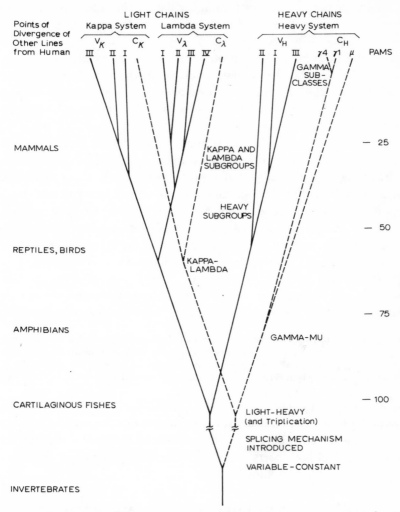

Fig. 22. Evolution of human immunoglobulins. The six sets of genes ($V_\varkappa$, V_λ, V_H, $C_\varkappa$, C_λ, C_H) which code for human immunoglobulins have arisen from a single primordial gene present in a pre-vertebrate ancestor. Scale to the right: amount of change in accepted point mutations per 100 links (PAM), in each of two immunoglobulin chains since their divergence from a common ancestor. The points of divergence of the evolutionary lines leading to other vertebrate groups from that leading to man are shown at the left. (*Atlas*[10], Fig. 4-1, p. 33)

several polypeptides. The most primitive vertebrate in which the immuno-globulin has been studied with respect to the association of the constituent polypeptides is the sea lamprey, in which it was found by Penn et al.[150] that no disulphide bonds were involved in the linking that was entirely due to non-covalent bonds. The H and L chains in one of the types of λ_A immuno-
—globulin of the rabbit are held together by non-covalent bonds. Grey[148] suggests that the primitive V_1C_1 proteins had only weak interaction affinities, like light chains, and that epigenesis developed this non-covalent interaction to which the action of interchain disulphide bonds was added later. A mammalian immunoglobulin (for example, the 7S γ_G-immunoglobulin) has the structure schematized in Fig. 20.

Two pairs of polypeptide chains are held together by S–S bonds and also by non-covalent interactions. When digested with a proteolytic enzyme such as papain the immunoglobulin splits into fragments. Fc (crystallizable fragment) is formed by the C-terminal halves of both heavy chains. (Fc is responsible for complement fixation and tissue binding.) Two Fab (antigen-binding) fragments are liberated as a result of proteolysis by papain. Each of these Fab is composed of a light chain (L) and the N-terminal half of the heavy chain (Fd) (H). Disulphide bonds situated at the C-terminal end of the light chains bind this chain to the heavy chain. It is generally agreed that both light and heavy chains contribute to the combining sites. From their study of light chain evolution, Hood et al.[141] have concluded that:

"An inescapable conclusion is, therefore, that each light chain is encoded by two genes ....which are expressed as a single, continuous polypeptide chain. Thus the "one gene, one polypeptide chain" hypothesis must be broadened to include the special case of antibody light chains, which need "two genes, one polypeptide chain.""

Encoded by separate genes, the V and C regions of light chains therefore have separate pathways in molecular evolution but they function together to make a light chain and must have recognition biosemes. Evidence is available for the homology of both parts of a light chain. The evolution of the heavy chain still awaits further knowledge.

C. EXTENSIONS ON THE PATHWAYS OF AMINO ACID BIOSYNTHESIS

11. Extensions on the pathway of arginine biosynthesis

(a) Introduction

The pathway of the biosynthesis of arginine is represented in Fig. 23. This

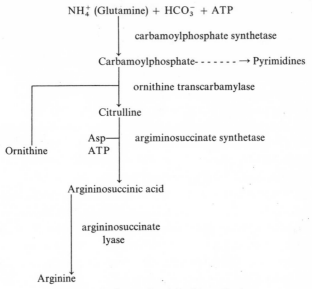

NH$_4^+$ (Glutamine) + HCO$_3^-$ + ATP

carbamoylphosphate synthetase

Carbamoylphosphate- - - - - - → Pyrimidines

ornithine transcarbamylase

Citrulline

Asp— argiminosuccinate synthetase

Ornithine   ATP

Argininosuccinic acid

argininosuccinate
lyase

Arginine

Fig. 23. Pathway of arginine biosynthesis.

pathway operates in the adult mammal; but developing mammals require arginine. The pathway belongs to the category of the primary catenary biosyntagms of biosynthesis and it is found in both bacteria and mammals. However, the arginine pathway has been lost in birds (Tamir and Ratner[151]), uricotelic reptiles (Brown and Cohen[152]), and in insects (Reddy and Campbell[153,154]). This pathway has been the subject of terminal extensions in evolution.

(b) Ureogenesis

This terminal extension is the result of the addition of arginase. It was once believed that the presence of this enzyme was evidence of ureogenesis. This view has been corrected and it now appears that the production of urea is conditioned not only by the presence of arginase but also by that of an operating arginine biosynthetic pathway. In the absence of this pathway arginase may contribute to the degradation of exogenous arginine. The presence of urease in a tissue may, however, break down the urea derived from endogenous or exogenous arginine. The presence of carbamoylphosphate synthetase does not imply the existence of the system of arginine biosynthesis, as this enzyme is also part of the catenary biosyntagm of the

biosynthesis of pyrimidine, through carbamoylphosphate and carba-
moylaspartate.

It was long believed that the nature of the end-products of amino acid
metabolism depends, in animals, upon the relative abundance of water in
embryonic development, a view that linked the nature of nitrogen catabolism
to ecological aspects of embryonic life. The theory was also related to the
toxicity of ammonia and it was generally believed that, while aquatic animals
have a large reservoir in which to excrete the ammonia resulting from
deaminations, animals which have been able to develop a terrestrial ecology
can only do this if they have acquired the means to dispose of ammonia
by "excretion synthesis" producing urea or uric acid.

The study of ammonemia is complicated by the formation of ammonia in
shed blood. As soon as the blood of a mammal leaves the circulatory system,
a rapid production of ammonia (ammoniogenesis α) occurs, lasting for about
five minutes (Conway and Cooke[155]). This is followed by a further small but
progressive rise. When the curve of ammoniogenesis α is extrapolated to zero
time, one can calculate the levels of ammonia *in vivo*. Conway and Cooke
have reported that in mammals and birds the curve reveals that the initial
blood ammonia level is extremely low (maximum value : 0.01 mg per 100 ml).

A similar result is found in poikilotherm vertebrates, such as frog, *Emys
orbicularis*, *Clemmys leprosa*, tench and trout (Florkin[156]). In vertebrates,
therefore, the levels of blood ammonia appear to be almost negligible. This
is also the case in insects, but not in the snail, lobster or crayfish. In *Helix
pomatia*, for instance, the blood contains from 0.7 to 2.0 mg of ammonia per
100 ml and in *Homarus vulgaris*, from 1.6 to 1.8 mg (Florkin and Renwart[157])
while the value is 1.9 in crayfish (Florkin and Frappez[158]). The notion of
circulating ammonia has also been confirmed in *Octopus* (Delaunay[159];
Potts[160]). These results show that it is not the toxicity of ammonia alone
which governs the general nature of excretory nitrogen products.

The whole catenary biosyntagm of ureogenesis has been identified as
homologous (sequence of identical enzymes) in the generality of animal taxa.

In Platyhelminthes, for example, many species are ureotelic due to ureogenesis and the same applies to *Fasciola hepatica* (Ehrlich *et al.*[161];
Kurelec[162]). In other species, the biosyntagm is either absent or non-
operative; this is the case in several digeneans (Goil[163]) and in several
aquatic turbellarians (Campbell and Lee[164]).

When cultured *in vitro*, adult parasitic nematodes excrete substantial
amounts of urea (4–15 percent of total nitrogen). As was shown by Cavier

and Savel[165], *Ascaris lumbricoides* excretes 7 per cent of its nitrogen excreted in the form of urea when bathed on its surface by water, but when the animal is put in a U tube in such a way that only the free ends are in contact with saline, the proportion of nitrogen excreted as urea rises to 51 per cent. This is an example of facultative ureotelism resulting from insufficient contact with water at the body surface.

That urea is produced in nematodes through the catenary ureogenetic biosyntagm was substantiated by the fact that citrulline and ornithine were found to increase urea production, and by the detection of arginase in, for example. *Ascaridia galli*, *Nematodirus* spp., and *Ascaris lumbricoides* (W. P. Rogers[166]; Cavier and Savel[165]). Ureogenesis appears to have been lost in a number of nematodes such as *Ditylenchus triformis*, *Trichinella spiralis*, *Nippostrongylus brasiliensis*, *Caenorhabditis briggsae*, etc.

Although normally ammoniotelic, the earthworm *Lumbricus terrestris* excretes as much as 90 per cent of its non-protein nitrogen in the form of urea during starvation (Cohen and Lewis[165]; Needham[168]). The complete array of enzymes of the ureogenesis catenary biosyntagm, insuring *de novo* synthesis of arginine and urea from CO_2 and NH_3 has been demonstrated in the earthworm gut tissue. It has also been shown that they function *in vivo* to synthesize *de novo* both arginine and urea (Bishop and Campbell[169]). During starvation, the earthworm becomes truly ureotelic. The increase of urea excretion is due to an increase in the level of the urea-cycle enzymes. As they are normally ammoniotelic, earthworms do not seem to be adversely influenced by ammonia. The amount of urea excreted seems to depend upon the amount of metabolic water available since inanition increases the excretion of urea. The importance of the shift from the ammoniotelic to the ureotelic metabolism appears to lie in osmoregulatory aspects.

In the case of molluscs, there is ample evidence of the presence of the whole catenary biosyntagm of ureogenesis in gastropods (see Campbell and Bishop[170]), a unique example among uricotelic animals. Any arginine formed by the tissues of *Otala lactea* and *Helix pomatia* is transformed into urea which is broken down into CO_2 and NH_3 due to the presence of urease (Speeg and Campbell[171]).

It was believed for some time that the teleosts only had arginase (Brown and Cohen [172]). But more precise analysis showed that a full series of ureogenesis enzymes could be detected in many fresh-water and marine teleosts, though at small activity levels (Huggins *et al.*[173]), which accounts for the ammoniotelism of teleost fishes.

References p. 220

The liver of the chimaeroid *Hydrolagus colliei* (L. J. Read[174]) and of the lungfish *Protopterus aethiopicus* (Janssens and Cohen[175]) also contains the whole outfit, as does the liver of the coelacanth *Latimeria chalumnae* (Goldstein *et al.*[176]), and of elasmobranches (Baldwin[177]; Brown and Cohen[172]; Anderson and Jones[178]). In the liver of the cyclostome *Entosphenus tridentatus* (Read[179]) only arginase and carbamoylphosphate synthetase were detected. As cyclostomes excrete almost nothing but ammonia, it appears that they use carbamoylphosphate synthetase in pyrimidine biosynthesis and not in ureogenesis. The ureogenesis enzymes have been shown to occur in the liver of all Amphibia and it is of particular interest that the levels of carbamoylphosphate synthetase reflect the extent of ontogenetic development. The shift from ammonia to urea depends on some type of induction of the urea cycle at metamorphosis during which an enhanced level of all enzymes is observed (literature in Balinsky[180]).

Chelonian reptiles are extremely versatile and have the ability to excrete their nitrogen as ammonia, urea and/or uric acid depending upon species and environment. They seem to have all the components for ureogenesis, whereas in the liver of lizards and snakes the full complement has not been found (literature in Coulson and Hernandez[67]). This also applies to birds. It is sometimes claimed, from the graph of nitrogen excretion in the chick, that during embryonic development the bird passes through a biochemical recapitulation, excreting ammonia first, then urea and finally uric acid. Needham *et al.*[181] have shown that the urea excreted by birds derives from the effect of arginase on the arginine derived from the proteins of the yolk.

Mammals are fully equipped with ureogenesis enzymes. It appears from the knowledge we now possess that the potential for biosynthesizing the whole array of ureogenesis enzymes has been repeatedly activated in the course of evolution. There are cases, for example in Amphibia, of a transition to a high amount of synthesized enzymes accompanying a transition from larval to adult conditions. The implementation of the pathway may also result from changes in the environment as in the case of *Ascaris*.

It is often stated that in particular forms the enzymes are deleted. As Watts and Watts[182] emphasize, this view always requires caution, as the example of teleost fishes testifies.

(c)   The meaning of ureogenesis

As stated above, it was once maintained that the nature of the end product of amino acid catabolism was related to the abundance of water during the animal's embryonic development. Ureogenesis was thought to prevent —in the case of the cleidoic* eggs—poisoning by ammonia.

It nevertheless appears that the presence of the ureogenesis pathway is better accounted for as the means of solving various kinds of osmotic problems resulting from environmental factors. In the case of selachians for instance, in the transition from fresh water to sea water in evolution the problem of the equilibration with the exterior salty medium was solved through reliance upon a pathway repeatedly called upon in phylogenic ancestry, viz. that of ureogenesis. The latter was used in this particular case to insure equilibrium with the medium by conserving a high concentration of small nitrogenous molecules (urea) in the body fluids as well as in the tissues.

Smith[183] has pointed out that there is much more justification for considering the cleidoic egg to be an adaptation secondary to urea retention. Having developed synthesis and retention in the adult as a valuable osmotic asset, these animals

"have found it convenient to enclose the embryo within a closed egg until such time as the membranes of the developing embryo can function to restrain the diffusion of the maternal gift of urea, as well as the complement added by its own metabolism. This view explains why, as an alternative to the cleidoic egg, the Elasmobranchii have resorted to viviparous reproduction. Intra-uterine development will serve the purpose of protecting the young embryo against loss of urea better even than an egg-case"[183].

It is understood that in Amphibia metamorphosing from a water larva to a terrestrial adult, the advent of ureogenesis in the latter remedies the loss of the watery surrounding entailing a rapid loss of ammonia, and replaces ammonia by the less toxic urea. But this does not exhaust the physiological radiations of the acquisition of ureogenesis. The discovery in Thailand by Neill[184] of a marine frog, Rana cancrivora, has raised the number of osmoregulatory uses of the ureogenesis pathway. Rana cancrivora has been the subject of a number of studies by Gordon et al.[185]; Schmidt-Nielsen and Lee[186], and Thesleff and Schmidt-Nielsen[187]. In this species the concentration of the blood follows the variations of the external medium and it has been observed that the inorganic concentration of blood in sea water is twice that in fresh water, the osmotic deficit being compensated by a great amount of urea whose concentration may reach 480 mM.

---

* enclosed within a protective shell.

Bufo viridis adapts to different concentrations in the surrounding medium. As was shown by Schoffeniels and Tercafs[188], when *Bufo viridis* is placed in solutions more concentrated than fresh water the urea concentration in the blood as well as that of inorganic constituents is increased, whereas in *Rana temporaria* and in *Bufo bufo*, only the latter are increased. If we compare the results obtained with *Rana temporaria*, *Bufo bufo* and *Bufo viridis* (including the data obtained by Gordon[189] on this species), as well as those of Gordon *et al.*[185] on *Rana cancrivora*, and of Scheer and Markel[190] on *Rana pipiens*, it appears that species able to live in salt solutions containing more than 15 g NaCl per liter have a higher urea concentration (a physiological uremia) in fresh water, and that they are able to increase the blood urea concentration when transferred to more concentrated media.

This adjustment in the form of increased uremia also occurs in the case of dehydration. *Xenopus laevis* exhibits, in dehydration, increased uremia up to ten times the normal value (Balinsky *et al.*[191]). Urea also plays a part in the osmoregulation of *Testudo hermanni* during seasonal changes (Gilles-Baillien and Schoffeniels[192]). The compensating effect of uremia in dehydration is also observed in mammals. In the dromedary, for instance, dehydration is accompanied by increased uremia (Charnot[193]) and the same is also observed in man when primary dehydration sets in (see Florkin[194]).

(d)   A terminal extension on the primary biosyntagm of arginine biosynthesis

All the presently known phosphagens are found in Annelida and they are, in the case of each of them, homologous (secondary homology) with the same phosphagen of other animal phyla though different phosphagens are not always homologous. It is therefore convenient to consider this aspect of biochemical evolution in the context of the phylum which is the richest in extensions of that kind, the Annelida (Fig. 24). The fact that all known phosphagens are found in this phylum contrasts with the uniformity of Arthropoda or Mollusca—all of which have only phosphoarginine—or vertebrates which have phosphocreatine only. The diversity of annelids in this respect is a character of their phylum and class characteristics may be ranked among the patterns of endpoints of the biosynthetic pathways.

Fig. 24. Pathways of biosynthesis of guanidines and phosphagens. (A) Pathways to phosphocreatine and phosphoglycocyamine; (B) Pathways to phospho-opheline and phospholombricine; (C) Pathways to phosphotaurocyamine and phosphohypotaurocyamine; (D) Pathways to phosphoarginine and to other guanidines derived from, or amidinated by arginine. Uncertain steps are shown by broken arrows. (Needham[195])

A.

Glycine $\xrightarrow{\quad + H_2N-\overset{\overset{\displaystyle NH}{\|}}{C}-(arginine)\quad}$ Glycocyamine $\xrightarrow{\quad ATP\quad ADP\quad}$ | Phosphoglycocyamine |

$\downarrow$ +CH$_3$ (methionine)

Creatine $\xrightarrow{\quad ATP\quad ADP\quad}$ | Phosphocreatine |

B.

Ethanolamine $\xleftarrow{\quad -COOH\quad}$ Serine

$\downarrow$ + PO$_4$

Ethanolamine Phosphate $\xdashrightarrow{\quad + H_2N-\overset{\overset{\displaystyle NH}{\|}}{C}-?\quad}$ Guanidinoethanol Phosphate $\xrightarrow{\quad +CH_3\quad}$ Opheline $\xrightarrow{\quad ATP\quad ADP\quad}$ | Phosphoopheline |

$\downarrow$ + D - Serine

D-serine ethanolamine phosphate $\xrightarrow{\quad + H_2N-\overset{\overset{\displaystyle NH}{\|}}{C}-(arginine)\quad}$ Lombricine $\xrightarrow{\quad ATP\quad ADP\quad}$ | Phospholombricine |

C.

Cysteine sulphinic acid $\xleftarrow{\quad Oxidation\quad}$ Cysteine $\xleftarrow{\quad +SH\ (methionine)\quad}$ $\leftarrow$ $\leftarrow$ Serine

$\downarrow$ -COOH

Hypotaurine $\xrightarrow{\quad + H_2N-\overset{\overset{\displaystyle NH}{\|}}{C}-(arginine)\quad}$ Hypotaurocyamine $\xrightarrow{\quad ATP\quad ADP\quad}$ | Phosphohypotaurocyamine |

$\downarrow$ Oxidation $\qquad\qquad$ $\downarrow$ Oxidation

Taurine $\xrightarrow{\quad + H_2N-\overset{\overset{\displaystyle NH}{\|}}{C}-(arginine)\quad}$ Taurocyamine $\xrightarrow{\quad ATP\quad ADP\quad}$ | Phosphotaurocyamine |

D.

HCO$_3^-$ + NH$_4^+$ + ATP

Citrulline $\quad$ P$_i$

Aspartate

Carbamyl phosphate $\quad$ ADP

Argininosuccinate $\quad$ Ornithine

$\quad$ Urea

Fumarate $\quad$ ATP ADP

Protein $\leftarrow$ Arginine $\rightarrow$ | Phosphoarginine |

+Pyruvate $\quad$ Putrescine

Octopine

Spermidine $\quad\quad$ $+ H_2N-\overset{\overset{\displaystyle NH}{\|}}{C}-(arginine)$

$+ H_2N-\overset{\overset{\displaystyle NH}{\|}}{C}-$ (Arginine) $\quad$ Agmatine

Monoamidinospermidine I $\quad\quad$ $+ H_2N-\overset{\overset{\displaystyle NH}{\|}}{C}-$

$\quad\quad$ Arcaine

$+ H_2N-\overset{\overset{\displaystyle NH}{\|}}{C}-$ (arginine)

Hirudonine

References p. 220

12. Extensions on the pathways of the biosynthesis of aromatic amino acids

(a) Introduction

In bacteria, molds and plants, the biosynthesis of the aromatic amino acids is accomplished *via* the non-oxidative pathway going through shikimic acid and diverging into two branches, one proceeding towards phenylalanine and tryptophan, the other towards tyrosine. The shikimic pathway, leading in plants to the biosynthesis of coumarins, lignins, melanins and a number of alkaloids does not exist in animals. The phenylalanine of their diet, originating from the shikimic pathway, is used by them in a second episode in a novel, oxidative pathway leading to tyrosine as well as to epinephrine, thyroxine, tanning agents and animal melanins.

(b) Derivatives of phenylalanine and tyrosine (see Fig. 25)

(i) Tanning agents in Arthropoda

The phylum Arthropoda is the largest of the animal kingdom. Arthropods differ from their annelid ancestors in a number of characteristics, the most striking of which is their articulated exoskeleton. When thin and little sclerotized, as in caterpillars or maggots, the cuticle bends freely but when thickened by sclerotization or additional impregnations with calcium carbonate, an exoskeleton characterized by stiff sclerites separated by joints permits a perfecting of muscles for many purposes including flight as in pterygote insects. This is one of the main reasons for the great evolutionary success of Arthropoda. This exoskeleton contains chitin in association with proteins, but chitin is also an important constituent of molluscan shells and furthermore, it is generally confined to the procuticle. The essential factor in the rigidity of the cuticle of Arthropods lies in the process of hardening by sclerotization. The nature of the exoskeleton implies the alternation of periods of growth and of shedding of cuticular structure (molting).

Sclerotization of cuticular proteins of arthropods involves the action of phenoloxidases on phenolic substrates.

If we take the example of the last instar of the blowfly larva, we see that early in the instar, tyrosine undergoes transamination, giving 4-hydroxy-phenylpyruvic acid as a prelude to a conversion to lactic and propionic derivatives. At the last stage of the instar, tyrosine is in its greater proportion transformed by hydroxylation into dopa, which is decarboxylated to

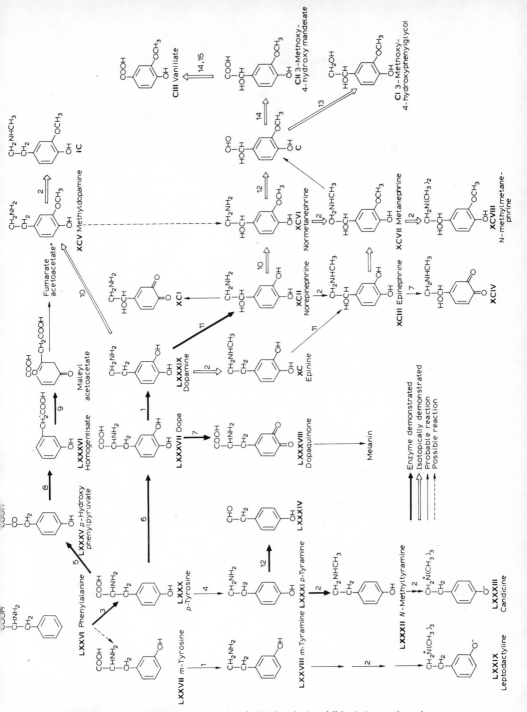

Fig. 25. Metabolism of phenylalanine and tyrosine derivatives in Amphibia. 1, Aromatic amino acid decarboxylase; 2, phenylethanolamine methyltransferase; 3. phenylalanine hydroxylase; 4, tyrosine decarboxylase; 5, tyrosine transaminase; 6, tyrosine hydroxylase; 7, DOPA oxygenase; 8, hydroxyphenylpyruvate hydroxylase (inactive form found); 9, homogentisate oxygenase; 10, catechol *O*-methyltransferase; 11, dopamine hydroxylase; 12, monoamine oxidase; 13, reduction: 14. oxidation; 15, oxidative decarboxylation. (Balinsky[180])

dopamine and acetylated to give N-acetyldopamine. The hormone ecdysone shifts the metabolism towards the new form by inducing the formation of dopa-carboxylase, catalyzing the conversion of dopa to dopamine. A phenoloxidase catalyses the hydroxylation of tyrosine and a transacetylase the acetylation of dopamine.

The same series of reactions has been observed in a crustacean, the cockroach (Mills *et al.*[196]). In the process of sclerotization (tanning) the cuticular protein reacts with quinone to give structures cross-linked with covalent bonds. The quinone has been identified in a very few cases.

(ii) Iodinated thyronines

In the evolution of vertebrates the nature of their thyroid hormone (composed of thyroxine and of triodothyronine) has not varied.

As the thyroid of the lamprey develops out of the endostyle of its larva, it was natural to enquire into the homologous tissues of the endostyles of protochordates (Urochordata and Cephalochordata) where organic bound iodine was identified. In the endostyle of protochordates, thyroid hormones are certainly present[197-200]. This lends support to the view that thyroidal biosynthesis appeared before vertebrates.

The binding of iodine to tyrosine is common in invertebrates at sites where structural proteins are laid down in the presence of an oxidase system able to promote the oxidation of iodide to iodine[201,202]. But at these sites the iodinated tyrosines are not coupled to form iodothyronines in significant amounts[203,204]. At other sites, nevertheless, the presence of monoiodotyrosine, diiodotyrosine and triiodothyronine has been demonstrated, for instance in the mucus layer of a nemertine[205]. The thyroid is derived from the endostyle which is not an endocrine organ but a digestive one, producing proteins which form the filtering membranes used in ciliary feeding. Therefore, iodine-binding "may have appeared in the endostyle initially as a metabolic by-product of its protein metabolism" (Barrington[206]).

In this context it is important to note that in Cephalochordata such as *Branchiostoma (Amphioxus) lanceolatum*, iodine concentrates in the endostyle. Here the secretion takes on the holocrine mode, the endostyle cells in which the hormones are synthesized going toward the digestive tube where they are hydrolyzed and where the thyroid hormones are recovered by intestinal absorption. In larval cyclostomes the same mode prevails, but when they reach the adult stage the thyroid hormones are liberated by merocrine secretion oriented towards the circulating blood. This fate would have been the consequence of the presence of any thyroxine present in the secretion discharged in the intestinal tract by an endostyle as a by-product of its protein metabolism, and become the starting point of an endocrine fortune of iodinated derivatives of tyrosine.

(iii) From phenylalanine and tyrosine to phenylalkylamines
  This pathway is the catenary biosyntagm of catecholamine biosynthesis, among a number of products.
  The catecholamines appear to have, late in evolution, taken the role of synaptic transmitter (in vertebrates) assumed by serotonin in worms, molluscs and some crustaceans (Bacq and Florkin[207]). Noradrenaline has not been found in coelenterates, echinoderms, crustaceans, molluscs (with the exception of cephalopods) or ascidians (see Euler[208]; Bacq and Florkin[207]) but it has been known for a long time that adrenaline (or noradrenaline) is found in certain neurones of the ganglia in annelids (Gaskell[209]). In the brain of amphibians, the catecholamines controlling the brain activity are found largely as adrenaline (epinephrine), while in other vertebrates it is mainly noradrenaline (norepinephrine). Some insects biosynthesize dopamine, noradrenaline and adrenaline (Östlund[210]). Dopamine, a precursor of adrenaline, probably acts as a neurohumoral regulator in the brain of mammals (Bertler and Rosengren[211]) as well as in the brain of *Helix* (Kerkut and Walker[212]). It plays also a regulatory role in the heart of *Venus mercenaria* (Greenberg[213]). Dopamine is the only catecholamine identified in the ganglia of pelecypod or gastropod molluscs (Sweeney[214]) which is consistent with the view that there it plays the role assumed in other forms by adrenaline or noradrenaline.

(c) Tryptophan derivatives

(i) From tryptophan to serotonin (5-hydroxytryptamine, 5-HT) and other indolealkylamines
  Serotonin, in the skin of amphibians, appears as an allomone playing the role of a repellant. In *Bufo* species a number of other repellants derived from serotonin are found. These include different stages of serotonin N-methyla-tion (N-methylserotonin, bufotenine, bufotenidine, etc.) (see Fig. 26).
  In amphibian skins, O-sulphate esters of these compounds are also biosynthesized (bufothionine, bufoviridine, etc.), as well as O-methylated ones (N-methyl-5-methoxytryptamine, O-methylbufotenine, etc.).

References p. 220

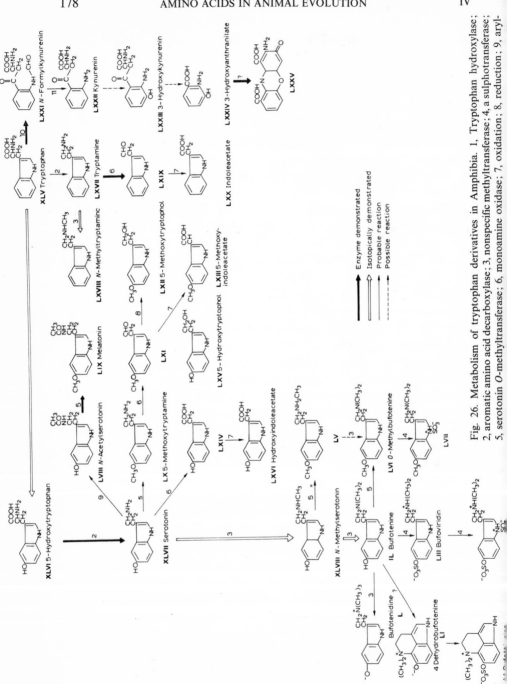

Fig. 26. Metabolism of tryptophan derivatives in Amphibia. 1, Tryptophan hydroxylase; 2, aromatic amino acid decarboxylase; 3, nonspecific methyltransferase; 4, a sulphotransferase; 5, serotonin O-methyltransferase; 6, monoamine oxidase; 7, oxidation; 8, reduction; 9, aryl-

5-Hydroxytryptamine is present at a high concentration in the nervous system of annelids, molluscs and some crustaceans in which it has taken the function of transmitter at the nervous synapse level (see Bacq and Florkin[207]).

(ii) From tryptophan to ommochromes

These pigments were first discovered in the eyes of insects but they do not act as visual pigments. Some have a low molecular weight (ommatins) while other have a high one (ommins).

The signified of ommochromes remains unknown (literature in Kennedy[215]).

D. PHYSIOLOGICAL RADIATIONS OF FREE AMINO ACIDS

13. Free amino acids as compensatory osmotic effectors in marine invertebrate tissues

It is known that the first Metazoa evolved in the sea. One of the pathways of evolution leads to the problem of the persistence of life when organisms invaded other media.

As far back as 1900, the relationship between osmolar concentration, inorganic constituents' concentration and organic constituents' concentration has been clearly defined by L. Fredericq[216] in the most simple of ways. He tasted the blood of a lobster and of other marine invertebrates, and found it as salty as sea water. He tasted the blood of a number of marine fishes and found it less salty than sea water but as salty as the blood of freshwater fishes. This was also true of the blood of a crayfish.

He then used the cryoscope to determine the osmolar concentration of the different types of blood and of the juice extracted from the tissues of the various animals; he also determined the amount of ash in the different samples. He explained the discrepancy between osmolar concentration determined cryoscopically and ash content by postulating the presence of important amounts of small organic molecules in the tissues of marine invertebrates and in the blood and tissues of elasmobranch fishes. The latter were identified as urea by Rodier[217] while the nature of the small organic molecules accounting in large part for the osmolar concentration of the tissues of marine invertebrates (free amino acids, taurine, trimethylaminoxide, etc.) was recognized only half a century later. A paper by Camien *et al.*[218] started this new development.

In lobster muscle, these authors found that the concentration of the 15 free amino acids they determined amounted to 440 mmoles per liter of intra-cellular water, while the concentration of chloride amounted to 450 per 1000 g of serum. In the tissues of the lobster, therefore, the free amino acids are major osmolar effectors; their contribution corresponds to half the value of the molecular concentration. The same amino acids account for 360 mOsmoles per liter of water in the muscles of the annelid *Perinereis cultrifera*[219], 240 in the muscles of *Nereis diversicolor*[219], 430 in the muscles of *Arenicola marina*[220], 330 in the gastric coeca of *Asterias rubens*[221], 300 in the muscles of *Leander squilla*[222], 400 in the muscles of *Carcinus maenas*[223], and 350 in the muscles of *Eriocheir sinensis*[224] adapted to sea water. The tissues of the marine species mentioned are approximately in osmotic equilib-rium with sea water (± 1000 mOsmoles per liter).

The same amino acids account for 170 mOsmoles in the muscle of *Astacus astacus*[225], the tissues of which are in osmotic equilibrium with blood containing 430 mOsmoles per liter. The existence of low values for the osmolar concentration of the free amino acid pool of tissues was also observed in *Anodonta cygnea* compared to *Mytilus edulis* and to *Ostrea edulis*, and in *Hirudo medicinalis* compared to *Sipunculus nudus*[226]. It appears that in spite of the existence of an anisosmotic extracellular regulation in fresh-water forms, the concentration of the inorganic constituents of the internal medium is always lower than in the marine forms. The equilibrium has been maintained in the cells by a lowering of the organic constituents. However, while the tissues of invertebrates contain variable amounts of free amino acids the latter are present in low concentrations in the body fluids of animals (with the exception of insects).

The adaptation to terrestrial life has also led to a lowering of the con-centration of the inorganic constituents of the internal medium and the equilibrium between tissues and blood has been accomplished through a decrease in the amount of compensating small organic molecules present in cells. In the muscles of vertebrates, the value of the total concentration of these organic components (203–224 mg per 100 g of fresh tissue in the rabbit[227], 345 in the cat[228], 223 in the chicken[227]) contrasts sharply with that found in the lobster (3000 mg per 100 g of fresh tissue[228]). But this does not result in a radical change in cell composition with respect to the inorganic constituents which appear to be preserved by the adaptive mechanism just described (adjustment of the pool of small organic molecules). Instead of a constancy of the internal medium with respect to its inorganic constitution, which was at one time considered a necessity, what has remained the same is the inorganic component of *cells*.

The intracellular free amino acid pool of invertebrates is not systematically higher than that of vertebrates. For instance, the total concentration in the muscles of *Hirudo medicinalis* and of *Anodonta cygnea*[226] is lower than in all vertebrates so far studied.

In a definite set of conditions, the pattern of the intracellular amino acid pool is more or less characteristic of a given cell differentiation in a given species (Duchâteau and Florkin[227]). This contrasts with the fact, demonstrated by several investigators, that the amino acid composition of the whole protein component is not very different from one species to another (Beach *et al.*[229]; Block[230]). This is not unexpected. Each cell contains a large number of proteins different with respect to composition and amino acid sequence. The concentrations of these proteins are also different and they have different speeds of turnover. Consequently there is nothing astonishing in the fact that the total amino acid pattern of a cell can be different from that of the free amino acid pool of the same cell. In fresh-water and terrestrial species, the inorganic components of the body fluids are no longer determined by the composition of the surrounding sea water as in marine invertebrates. The ionic and anisosmotic extracellular regulation maintains the inorganic component of blood above that of the external medium, and the intracellular isosmotic state with respect to the blood is insured by a lowering of the concentration of the pool of intracellular small organic molecules. This, at least, is generally true. But in insects, the osmolar concentration which resulted from the inorganic constituents in the body fluids of their marine ancestors has been kept at a high level due to substitution of a part of the inorganic constituents by amino acids, whose level in insects is generally very high in hemolymph as well as in cells.

14. Free amino acids as osmolar constituents of hemolymph and tissues of insects

In the marine ancestors of Arthropoda, the osmolar concentration of body fluids was nearly equal to that of the surrounding sea water and essentially insured by inorganic constituents, mainly sodium and chloride. By contrast, the situation in cells was, as we have said, quite different.

In the apterygote *Petrobius maritimus*, the most primitive insect so far studied, the Na^+ and Cl^- ions together insure the osmolar concentration of body fluids (Lockwood and Croghan[231]) as in marine crustaceans.

In the phylogeny of insects, a tendency is observed toward replacement of the inorganic ions by small organic effectors (Sutcliffe[232,233]) including amino acids to keep the high osmolar concentration of the body fluids at the level of the osmolar concentration of cells (instead of reducing the osmolar concentration of cells by lowering the organic osmolar effectors, as happens in other phyla).

In the most primitive pterygotes (Paleoptera and the majority of Poly-neoptera) the sum of the inorganic ions amounts to almost half of the total osmolar concentration, while chloride ions remain the main anions. A rather small but not negligible portion of the osmolar concentration in the hemolymph is insured by organic molecules. This Sutcliffe considers the basic type of hemolymph in pterygote insects. A second, similar type is recognized in Phasmida, with magnesium replacing sodium in the cationic pattern and the inorganic phosphates increased.

In the most primitive of endopterygotes, the Oligoneoptera and Diptera (Fig. 27) a third type is found, characterized by a marked lowering of the concentration of chloride ions and an increase of the concentration of small organic molecules in the hemolymph. This increase in the osmolar participation of the small organic constituents appears to have arisen very early in the phylogeny of Oligoneoptera, the origin of endopterygotes being considered monophyletic.

The most specialized type of hemolymph structure corresponds to type IV of Sutcliffe[233]. It is found in Lepidoptera, Hymenoptera and many Coleoptera. In these orders, the importance of inorganic constituents as osmolar effectors in hemolymph is considerably diminished and the main osmolar effect is due to small organic molecules, including free amino acids. Coleoptera, Hymenoptera and Lepidoptera are considered to have been independently derived from distinct stocks of primitive Oligoneoptera and consequently the increase in small organic molecules appears to be of polyphyletic origin.

The increasing importance of free amino acids as osmolar effectors in insect hemolymph is expressed by the fact that their concentration in exopterygotes corresponds to values lower than 700 mg per 100 ml while it is higher than 700 mg per 100 ml in endopterygotes (Jeuniaux[234]). The free amino acid distribution between tissues and hemolymph of insects has been studied in Bombyx mori larvae (Bricteux-Grégoire and Florkin[235]) and

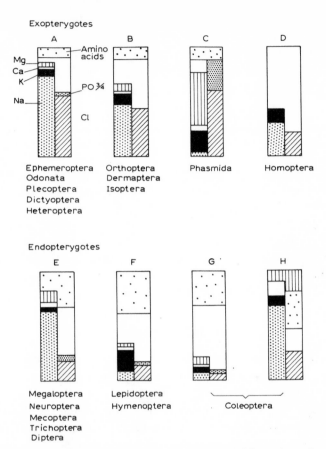

Fig. 27. Osmotic effects of components illustrated as percentages of the total osmolar concentration of hemolymph in pterygote insects. Each block in the figure is visualized as two vertical sections, each section representing 50% of the total osmolar concentration. The percentage contributions of cations are illustrated in the left-hand section, with sodium at the base (stippled), followed by potassium (black area), calcium (white area) and magnesium (vertical stripes). Anions are illustrated in the right-hand section, with chloride at the base (oblique stripes), followed by inorganic phosphate (fine stippling). Where possible, free amino acids are illustrated in equal proportions in both sections (coarse stippling). The large blank area in each block represents the proportion of the total osmolar concentration that must be accounted for by other components of the hemolymph. (Sutcliffe[233])

in the southern armyworm, *Prodenia eridania* (Levenbook[236]). In both species, the total concentration of free amino acids is about the same in cells in hemolymph.

References p. 220

15. Variations of the free amino acid pool of insect hemolymph during development, metamorphosis and cocoon spinning

The composition of the free amino acid pool of the hemolymph of insects is highly variable (Duchâteau and Florkin[237]); in the same species, different patterns are observed in different stages of development.

"The aminoacidemia of an insect species may be defined as being a succession of steady states expressed by a succession of patterns specific to the different instars of this species and to particular ecological or physiological events" (Florkin and Jeuniaux[238]).

The factors involved in the variations of concentration of a number of hemolymph free amino acids during the 4th larval molt and the eclosion of *Bombyx mori* have been reviewed in section 37 of Chapter I (Vol. 29 A), to which the reader is referred.

The variations of the free amino acid pool of the hemolymph of other insects have been observed during the development (*Prodenia eridania*, Levenbook[236]; *Philosamia ricini*, Pant and Agrawal[239]; *Antheraea pernyi*, Mansingh[240]; *Formica polyctena*, Brunnert[241]).

16. Diapause

As shown by Williams[242], a reaction of the organism of *Hyalophora cecropia* to external conditions is found when exposure to low temperature, which exerts an influence on the control of the suspension of nymphal diapause, is followed by exposure to a higher temperature. Due to the circulation of ecdysone released from the prothoracic glands under the influence of the brain hormone, the end of diapause is accompanied by increased respiration and a rise in respiratory pigment turnover and .concentration. It is also accompanied by a progressive pattern change of the aminoacidemia, ranging from one characterized by a high proportion of lysine, arginine, alanine, proline, histidine and total glutamic acid, to a pattern with two marked peaks (glutamic acid and tyrosine). The change in the steady state of each amino acid is due to the new conditions brought about by the circulation of ecdysone and the liberation and incorporation of the amino acids concerned. In the case of glycine, the turnover of the free amino acid from the hemolymph to the adult tissues is higher in nymphs of *Sphinx ligustri* during their progress towards adulthood than in nymphs still in diapause (Bricteux-Grégoire et al.[243]).

17. Free amino acids and intracellular isosmotic regulation in euryhaline invertebrates

(a) Introduction

The existence of a variable distribution of NADH between respiratory chains and glutamate synthesis has been revealed in the course of studies on isosmotic intracellular regulation in euryhaline invertebrates, *i.e.* those able to tolerate wide salinity ranges in contrast to the stenohalines which are restricted to narrow salinity ranges of the fluid environment. The possibility of a euryhaline form succeeding in aquatic media of different salinities depends in large part on its ability to regulate the molecular concentration of both its cells and body fluids.

When a euryhaline marine invertebrate is transferred to a more diluted medium, the ionic concentration of its blood may be kept at a concentration different (higher) than that of the environment. This anisosmotic extracellular regulation (which does not prevail in all euryhaline forms) is the result of the activity of the excretory organs and of an active uptake of ions (see Shaw[244]). The cells of marine invertebrates, in contrast with the body fluids, are rich in potassium and poor in sodium and chloride. The inorganic components account only for about one half of the total osmotic pressure of the tissues. As shown in 1951 by Camien *et al.*[218], free amino acids play a large role as osmolar effectors of the cells which are composed, besides the inorganic components, of amino acids, quaternary ammonium derivatives and organic phosphate compounds.

The amount of amino acids is much higher in marine than in fresh-water Crustacea (Camien *et al.*[218]; Duchâteau and Florkin[245]). The same may be said of quaternary ammonium derivatives (Gasteiger *et al.*[246]; Beers[247]). The important role played in the establishment of the intracellular osmotic pressure by amino acids led to a study in a euryhaline form of the free amino acid concentrations in tissues in a concentrated and in a diluted medium. This was done in 1955 on the Chinese crab *Eriocheir sinensis* (Duchâteau and Florkin[245]), using the microbiological method of amino acid determination. Bricteux-Grégoire *et al.*[224] used the more accurate method developed by Moore and Stein. Table XII shows that in *Eriocheir sinensis*, when adapted to fresh water, the inorganic constituents of the muscles represent about 40 per cent of the osmotically active constituents. The rest of the osmolar concentration is due to small organic molecules, 50 per cent of which is re-

TABLE XII

Intracellular osmotic effectors in *Eriocheir sinensis* adapted to fresh water and to seawater
(Bricteux-Grégoire et al.[224])

Intracellular osmotic effectors	Content (mOsm/liter of water) in:			
	Fresh water	Seawater	Fresh water	Seawater
Cl	76.0	153.1	44.6	166.9
Na	68.5	140.8	41.1	146.9
K	56.8	159.0	84.5	133.1
Ca	11.7	8.1	5.2	11.2
Mg	9.2	22.4	9.2	25.3
Total inorganic effectors	222.2	483.4	184.9	483.4
Ala	17.1	46.1	18.1	71.9
Arg	36.7	56.0	36.5	54.7
Asp (total)	5.4	12.2	3.6	11.7
Glu (total)	15.0	36.8	10.3	28.2
Gly	46.5	73.4	57.0	108.5
Ileu	1.4	4.6	1.0	3.2
Leu	2.2	6.1	1.7	5.4
Lys + his + X	9.6	21.7	14.3	18.5
Phe	0.0	tr.	0.0	tr.
Pro	18.2	37.3	4.7	23.7
Ser	5.2	7.6	2.6	6.3
Thr	4.4	17.2	4.4	15.3
Tyr	0.0	tr.	0.0	tr.
Val	0.0	8.1	0.0	6.9
Total amino acids determined	161.7	327.0	154.2	354.3
Taurine	14.1	13.6	20.5	27.7
Trimethylamine oxide	49.9	73.9	45.3	75.8
Betaine	9.5	6.9	25.7	21.0
Undetermined nitrogen	108.5	187.7	89.3	131.9
Total effectors determined	565.9	1092.5	520.0	1094.1
Calculated osmolar concentration $(\Delta/1.87) \times 1000$	588.0	1117.6	588.0	1117.6

presented by the amino acids determined (including taurine). In the crabs adapted to sea water, an increase is observed in the level of all osmotic effectors. The total concentration of the free amino acids is approximately doubled. As tissue hydration is only slightly modified when the crabs are transferred from a diluted to a more concentrated medium (Scholles[248]; Bricteux-Grégoire et al.[224]) it was proposed (Florkin[249,250])

"to consider the variation of the amino acid component resulting from a change in the medium concentration, as exerting an intracellular osmotic regulation".

Many recent studies performed in different laboratories have concerned the free amino acid content of the tissues of euryhaline invertebrates (those who do not die when transferred to media whose concentration differs from that of their natural aquatic medium) and they have consistently been found and observed to be reversible in euryhaline forms, whatever the phylum to which they belong.

(1) Crustacea: Carcinus maenas (Duchâteau and Florkin[251]; Shaw[252,253]; Duchâteau et al.[223]); Leander serratus, Leander squilla (Jeuniaux et al.[222]); Eriocheir sinensis (Duchâteau and Florkin[245], Bricteux-Grégoire et al.[224]); Astacus astacus (Duchâteau and Florkin[225]).

(2) Arachnomorphs: Limulus polyphemus (Bricteux-Grégoire et al.[254]).

(3) Mollusca: Mytilus edulis (Potts[255]; Lange[256]; Bricteux-Grégoire et al.[257]); Gryphea angulata (Bricteux-Grégoire et al.[258]); Ostrea edulis (Bricteux-Grégoire et al.[259]); Rangia cuneata (Allen[260]); Crassostrea virginica (Lynch and Wood[261]); Tegula funebralis (Peterson and Duerr[262]); Glicimeris glicimeris (Schoffeniels and Gilles[263]); and Mya arenaria (Virkar and Webb[264]).

(4) Annelids: Nereis diversicolor; Perinereis cultrifera (Jeuniaux et al.[219]) and Arenicola marina (Duchâteau et al.[220]).

(5) Echinoderms: Asterias rubens (Jeuniaux et al.[221]); Strongylocentrotus droebachiensis (Lange[265]) and Ophiactis arenosa (Stephens and Virkar[266]).

(6) Sipuncula: Phascolopsis gouldii (Virkar[267]).

It is interesting to note, with respect to the molecular factors of tissue osmoregulation, that several of the euryhaline species mentioned, depend on the isosmotic intracellular regulation only (which keeps the total osmolar concentration of cells near the modified osmolar concentration of the body fluids) and that they lack anisosmotic extracellular regulation (which keeps the inorganic constituents of blood at a higher or lower concentration than in the medium). This is true of *Asterias rubens* (Binyon[268]), *Arenicola marina* (Schlieper[269]; Beadle[270]) or *Mytilus edulis* (Potts[255]). When these species are transferred from sea water to diluted media in which they are able to survive, the inorganic composition of the blood follows the variation of the medium.

Despite the absence of anisosmotic extracellular regulation the population of *Asterias rubens* extends, in the Baltic sea, as far as Rugen Island, in whose vicinity salinity is 8 parts per 1000 (Brattsbröm[271]; Segerstråle[272]; Schlieper[273]). Binyon[268] has shown that in the North Sea the population of *Asterias rubens* can stand, for a limited time, a degree of salinity as low as 23 parts per 1000.

In all cases so far studied of euryhalinity in marine invertebrates, intracellular isosmotic regulation has been found to prevail whereas anisosmotic extracellular regulation, which relieves the intracellular mechanism of a part of its duty, is not always present. We must therefore consider intracellular isosmotic regulation to be the more primitive mechanism to which in species with more extended euryhalinity, anisosmotic extracellular regulation adds a new range of possibilities.

The acquisition, obviously polyphyletic, of euryhalinity by marine invertebrates is one of the main steps in animal evolution as it represented a condition for the colonisation of estuaries, of fresh water and of lands, which was at the origin of the geographic radiation of animal forms. The survival of an aquatic animal in media of modified salinity depends on the presence of different systems, some mechanical, others molecular, acting synergistically and defining the degree of euryhalinity of the species. In some molluscs, *e.g.* *Littorina littorea* and in *Purpura lapillus*, the perivisceral fluid acts as a buffer between the blood and the external medium when the organism is transferred to diluted sea water, and the operculum closes the shell opening, thus preventing a rapid change in the composition of the perivisceral fluid (Hoyaux[274]). Euryhalinity depends on a number of properties at both the molecular and at higher levels and even then it is clear that survival in diluted sea water is possible only if the animal possesses the molecular adaptations that enable it to keep not only the concentration and the composition of its blood in the adequate state, but its cell volume constant as well. There are species which manifest "experimental euryhalinity" in the laboratory but no "ecological euryhalinity" in nature. Such is *Maia squinado*, for instance.

(b) Mechanism of the participation of free amino acids in the isosmotic intracellular regulation of Crustacea

This subject has been studied mostly in euryhaline Crustacea. From a number of experimental data we must conclude that while all amino acids show changes in concentration beyond those to be expected on the basis of hydration changes, the principal factor lies in changes of intracellular concentrations of amino acids such as alanine, glycine, glutamic acid and proline, which have been shown to be non-essential amino acids in Crustacea (Gilles and Schoffeniels[275]). The idea comes to mind that variations in the intracellular pool of free amino acids could be quantitatively dependent on the turnover rate of some proteins. Experimental data do not confirm this hypothesis. When *Eriocheir* is transferred from fresh water to sea water, the increase in free alanine in muscles is accompanied by a corresponding increase in total alanine determined in hydrolysates of whole tissues, proteins included. The results also show that while free proline markedly increases, the amount of total proline in hydrolysates of whole muscles increases proportionally (Florkin *et al.*[276]). On the other hand, no change can be detected in the electrophoric pattern of proteins during the adaptation of *Eriocheir sinensis* to sea water (Florkin and Hamoir, unpublished).

Isosmotic intracellular regulation is also accompanied by a modification of nitrogen excretion which increases when a euryhaline species is transferred to a diluted medium but decreases during adaptation to a hypertonic medium. This has been demonstrated for *Carcinus maenas* (Needham[277]) and *Eriocheir sinensis* (Jeuniaux and Florkin[278]; Florkin *et al.*[276]). On the strength of these experimental data it is likely that the regulation of the intracellular pool of free amino acids depends on the acquisition of mechanisms controlling the relative speed of the anabolism and catabolism of the amino acids concerned. This does not of course imply that individual molecules of amino acids may not derive from the protein turnover. It is the level of the pool which is modified by the increase or decrease in amino acid metabolism.

This view is corroborated by experiments performed by Schoffeniels and his collaborators on isolated nerves of *Eriocheir*. Schoffeniels[279] has shown that the total amino nitrogen is higher in the isolated nerve of *Eriocheir* incubated in a saline whose concentration is that of an animal adapted to sea water than in an isolated nerve incubated in a saline corresponding to the blood concentration of an animal living in fresh water.

References p. 220

Results obtained more recently by Gilles and Schoffeniels[280] experimenting with the same material suggest that the regulation of osmolar concentration during osmotic stress involves at least two processes. The concentration of some amino acids such as tyrosine, phenylalanine, leucine, isoleucine, or valine appears to be regulated by a mechanism involving modifications of the permeability of the nerve membrane. But the concentration of the non-essential amino acids is regulated by a mechanism involving modification of their intracellular metabolic mechanism, their variation being unaccompanied by variations in the medium. As the content of nerve proteins does not vary significantly during this experimentation it appears that the variations derive from a synthesizing process and not from a modification of the steady state between free amino acids and proteins. In these experiments, ammonia did undergo a significant variation in the nerve as well as in the medium. Schoffeniels has suggested that it is the modification of the ionic concentration within the cells which, in euryhaline forms, controls the mechanism responsible for the modification of the amino acid level. Schoffeniels and his group have enquired into enzymes involved directly and indirectly in nitrogen metabolism, the activity of which would be modified by the ionic composition of the medium (literature in Schoffeniels[281]; Gilles and Schoffeniels[282]; Gilles[283]; Schoffeniels[284]; Schoffeniels and Gilles[263]).

The enzymatic system involved in the metabolism of amino acids has not been found to be influenced by NaCl concentration, either in euryhaline or in stenohaline organisms. On the contrary, the influence of NaCl concentration changes has been found to differ in euryhalines and stenohalines at the level of the enzymes controlling the extramitochondrial ratio NADH/NAD (lactate dehydrogenase) as well as the transfer of reducing equivalents originating in the cytoplasm to the mitochondrial respiratory chain (3-glycerophosphate dehydrogenase). It is therefore the fate of the reducing equivalents which issue from the central metabolic system of the cell that, according to this theory, determines the difference between euryhaline and stenohaline Crustacea. There is also an influence of the ionic composition on the activity of enzymes directly involved in the anabolism and in the catabolism of amino acids, and particularly on glutamate dehydrogenase which controls the entry of ammonia into the amino acid pool. Glutamate dehydrogenase, which is important in the synthesis of glutamate, is activated by an increase in ionic concentration whereas the activities of aspartate aminotransferase and of alanine aminotransferase are not modified and the activity of serine hydrolyase is decreased. As stated by Schoffeniels and Gilles[263]:

"It can therefore be tentatively concluded that an increase in ionic concentration induces an increase in the glutamate synthesis. Under conditions of increased ionic concentration, serine hydro-lyase is inhibited thus leading to a reduced deaminating activity. In turn, the increased amount of glutamate would induce an increase in the amount of the other amino acids, the activity of the aminotransferase being unaffected" (see Fig. 28).

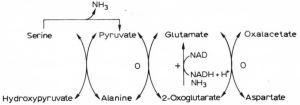

Fig. 28. Effect of NaCl on the activity of enzymes involved in the metabolism of amino acids. +, indicates an increase in enzyme activity; −, indicates a decrease; 0, indicates no effect. (Gilles[283])

This theory agrees with the observation that in euryhaline Crustacea, an increase in the amino acid pool and a decrease in nitrogen excretion is observed during adaptation to concentrated media (Needham[277]; Jeuniaux and Florkin[278]; Florkin et al.[276]). Schoffeniels concludes that when a euryhaline form of crustacean such as *Astacus* or *Carcinus* is transferred to a concentrated medium, a number of biochemical events which are observed (increase in free amino acid pool, decrease of nitrogen excretion and of oxygen consumption) can be explained at least partly, by the effect of changes in ionic concentration on the activity of enzyme systems related to amino acid metabolism.

The difference between euryhalines and stenohalines lies, according to these views, in the way in which the reducing equivalents carried by NADH are geared towards either amino acid metabolism or towards energy metabolism (Fig. 29).

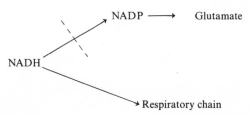

Fig. 29. Competition between glutamate synthesis and the respiratory chain. (Schoffeniels[284])

References p. 220

IV. DIVERSIFICATION IN ANIMALS AT THE LEVEL OF THE CENTRAL METABOLIC BIOSYNTAGM

A. INTRODUCTION

We may approach animal forms from the viewpoint of the diversification of a restricted number of catenary metabolic biosyntagms, such diversification resulting from terminal or lateral extensions which in turn arise from the acquisition of new tertiary or quaternary biosyntagms either by commutation or by insertion of already existing biomolecules. We find at the start the blueprint (DNA) of the genomic data for the synchronic epigenetic formation of the proteins involved, as well as the central metabolic catenary syntagm of cells, an ancestral feature which insures the availability of the free energy used in biosynthesis in the form of the energy-rich bonds of ATP. We may therefore begiñ with the central metabolic catenary biosyntagm and with examples of its diversifications in animal phylogeny.

B. LATERAL EXTENSIONS ON THE GLYCOLYTIC PATHWAY

18. Chitin pathway

(a) Chitin biosynthesis

Chitin biosynthesis appears as a lateral extension, located at the level of F-6-P, in the glycolytic pathway. The following schemes list the intermediates identified in insects by Candy and Kilby[285], and the demonstration by Jaworsky et al.[286] of the incorporation of uridine diphosphate N-acetyl-glucosamine into chitinodextrin.

$$\text{F-6-P} + \text{glutamine} \rightarrow \text{glucosamine-6-P} + \text{glutamic acid} \tag{1}$$

$$\text{Glucosamine-6-P} + \text{acetyl-CoA} \rightarrow N\text{-acetylglucosamine-6-P} + \text{CoA} \tag{2}$$

$$N\text{-Acetylglucosamine-6-P} + \text{UTP} \rightarrow \text{UDP–}N\text{-acetylglucosamine} + \text{PP}_i \tag{3}$$

$$\text{UDP–}N\text{-acetylglucosamine} + (\text{chitinodextrin})_n \rightarrow (\text{chitinodextrin})_{n+1} + \text{UDP} \tag{4}$$

(b) Regressive evolution of chitin biosynthesis

Whereas the protective barrier of the body of vertebrates is made of keratin, a complex of chitin and protein plays the same role in Arthropoda. Chitin biosynthesis is a primitive characteristic of animal cells, present at the level of the monocellular ancestor of Metazoa. Chitin, indeed, is the structural polysaccharide of a number of metaplasmatic membranes formed by Protozoa, such as the shells of Thecamoebia (= Testacida) and the cyst walls of some rhizopods and ciliates. The cells of the organisms which constituted the ancestral root of Porifera and diblastic acoelomates were probably equipped with chitin synthetase, since chitin is actually used by Spongillidae

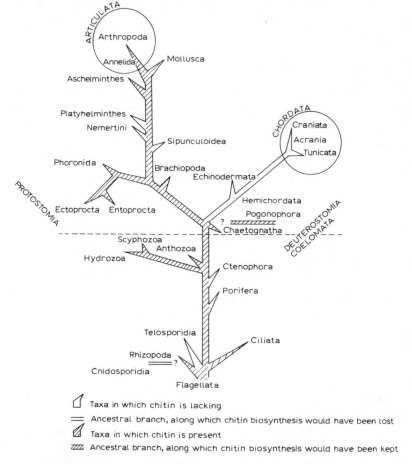

Fig. 30. Distribution of chitin, and the phylogenic relationship proposed by Marcus in 1958. (Jeuniaux[287])

(for the construction of the gemmule walls) and by many coelenterates, especially thecate and athecate Hydrozoa, for the building of peridermic structures (hydrorhiza, hydrocauli, hydrotheca). Chitin biosynthesis is extant in almost all Protostomia and is absent in three taxa only: Platyhelminthes, Nemertini and Sipunculoidea. As for Deuterostomia, chitin has been found in only two very primitive and aberrant of its taxa, *viz.* Chaetoptera and Pogonophera. Chitin biosynthesis and the presence of chitin synthetase is a primitive characteristic of animal cells, preserved during the evolution of the ancient forms as well as of the more advanced ones in the protostomian lineage, but lost at the start of the deuterostomian branch (Fig. 30). In Protostomia it is secondarily lost in a few groups such as Platyhelminthes and Nemertini.

(c) Physiological radiations of chitin

In the Protostomia which have retained chitin biosynthesis, the utilization of chitin is intensified; this is most notable in Arthropoda. Also a number of physiological radiations (at levels of integration higher than the molecular ones) are observed. Chitin is primarily used in the formation of protective envelopes, the rigidity of which is insured by tanning of a protein combined with chitin, by calcification resulting from the deposition of calcium carbonate, or by silicification. Such protective envelopes of organisms are represented by the theca and cyst walls of rhizopods and ciliates, by the hydrotheca of the calyptoblastic Hydrozoa, by the tubes of Phoronida and Pogonophora, by the shells of inarticulate Brachiopoda and of molluscs, and by the cuticles of Priapulida, Onychophora and Arthropoda. In Nematoda, chitin is used for the formation of one of the envelopes surrounding the eggs. It surrounds the latent form of life (gemmules) in *Spongilla*. Besides the function of protection of the individual, chitin in Hydrozoa and Bryozoa insures the skeletal unity of the colony. It contributes to the formation of locomotor organs, such as the locomotor appendages of Arthropoda and the rigid setae of Annelida. Various kinds of buoyancy organs are constructed with the use of chitin. Examples are the pneumatophores of Siphonophora, the shells of *Nautilus* and *Sepia* and the cuttlebone of *Sepia*. Chitin may also play a part in the adhesion of the organism to its substrate, as is the case in the cuticle of the peduncle and its ramifications in articulate Brachiopoda. At the level of the digestive function, chitin is also used in different structures. It contributes to the formation of several organs active in the capture or mastication of food (jaws and radula of gastropod and cephalopod molluscs). It con-

stitutes the gastric shield of pelecypods, which is in contact with the crystalline style producing enzymes acting extracellularly. Chitin is also used in the
protection of the mucosa of the mesentera and in the formation of faeces
(peritrophic membrane of Arthropoda).

Chitin is found either free or in combination with a protein (forming a
glycoprotein). Masked chitin (glucoprotein) dominates in the most primitive
groups (Protozoa, Porifera, Coelenterata) as well as in Arthropoda (free
chitin in the proportion of 5–17.5 per cent of total chitin) and in the setae of
polychetes (0.5–2 per cent of free chitin).

By contrast, the shells of molluscs contain chitinous layers (periostracal
and nacreous layers) characterized by a remarkably high proportion of free
chitin (25–68 per cent of total chitin).

(d) Regressive evolution in chitinolysis

Chitin is a homopolysaccharide, a homoglycan of linear structure. Mild
acid hydrolysis of chitin leads to the diholoside of N-acetyl-D-glucosamine
called chitobiose, as well as to the triholoside chitotriose. The enzyme
catalyzing the hydrolysis of chitobiose is called chitobiase, whereas the
enzymes catalyzing the hydrolysis of chitin are called chitinases. The
complete enzymatic hydrolysis of chitin leads to a complete conversion into
acetylglucosamine (Fig. 31). Small amounts of free acetylglucosamine can be

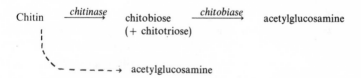

Fig. 31. The enzymatic hydrolysis of chitin.

liberated by the action of chitinase, but chitobiose and chitotriose account
for up to 99 per cent of the products. Chitinolysis is the result of the action of
two enzymes: chitinase and chitobiase. The ability of animal tissues to secrete
chitinase was long a controversial subject, but all doubt was finally removed
when the secretion of chitinase by isolated mucosa of reptiles was demonstrated (Dandrifosse[288]). Chitinase and chitobiase are synthesized by many
unicellular organisms such as bacteria and Protozoa, and are among the
first hydrolases to be secreted outside the cells by multicellular organisms

acquiring an extracellular mode of digestion.

In diblastic forms, the two layers inherit the same complete system as shown by Jeuniaux[74] but later on in evolution a specialization appears by regressive biochemical evolution of an enzymapheretic nature, involving the loss or repression of one of the two enzymes. Four cases were described by Jeuniaux[74].

(1) The case of the ectoderm of prostomians. There the chitobiase remains while the chitinase is lost, except in Nematoda and in the embryonic stages of Arthropoda. In these cases both enzymes are identified and they play an important role in the hatching of Nematoda and in the molting of Arthropoda.

(2) The case of the ectoderm of Deuterostoma, showing the complete loss of both enzymes.

(3) The case of the endoderm layer of Protostoma where both enzymes are conserved in all omnivorous forms and in those living on a diet containing chitin. On the other hand, in animals adapted to a diet deprived of chitin, such as the pigeon, the cow or the tortoise, both enzymes are lost.

(4) In insectivorous vertebrates, there is a loss of chitobiase, the chitinase being preserved and allowing for the penetration of the enzymes of the digestive juices into the mass of the prey.

19. Trehalose biosynthesis

It has been known for a long time that, with few exceptions, the reducing power of insect hemolymph is mainly due to non-saccharidic substances such as ascorbic acid, α-ketonic acids, uric acid, a number of reducing amino acids such as tyrosine, and other substances, still unidentified (Howden and Kilby[289]). The hemolymph contains no saccharose. It is known since the pioneer work done by Wyatt and Kalf[290,291] that a non-reducing dimer of α-glucose, trehalose, is the major blood sugar in most insects. In Table XIII the concentrations of a number of carbohydrates including trehalose are recorded. The concentrations of trehalose range from 202 mg/100 ml in the pupae of *Bombyx mori* to 4700–5200 mg/100 ml in the larvae of *Chalcophora mariana*, a buprestid. There are exceptions to this generalization. In *Phormia regina* trehalose is not found in the hemolymph of the larva, though it is found in pupae and adults (Evans and Dethier[292]; Wimer[293]). Trehalose is also lacking in the larvae of a number of Diptera, several species of Calliphoridae (Crompton and Birt[294]), *Sarcophaga bullata* and *Musca domestica*

TABLE XIII

CONCENTRATION OF TOTAL AND FERMENTABLE SUGARS

(expressed in glucose, mg/100 ml) and of glucose, fructose and trehalose in the hemolymph of insects (mg/100 ml)

(Florkin and Jeuniaux[238])

Species	Stage	Fermentable sugars (as glucose)	True glucose	Fructose	Trehalose
Dictyoptera					
Periplaneta americana	?	30	–	–	–
Leucophaea maderae	?	65	–	–	580–780
Orthoptera					
Schistocerca gregaria	Larvae	–	–	traces	800–1500
Coleoptera					
Hydrophilus piceus	Adults	5–31	–	–	300–500
Popillia japonica	Larvae	69	–	–	–
Chalcophora mariana	Larvae	–	–	–	4800–5300
Erbates faber	Larvae	–	–	–	3200
Hymenoptera					
Diprion hercyniae	Larvae	–	28	–	926
Apis mellifica	Adults	1000–4000	600–3200	200–1600	600–1200
Lepidoptera					
Phalera bucephala	Larvae	40	–	–	–
Prodenia eridania	Larvae	11	–	–	–
Bombyx mori	Larvae	9–28	1–3	1–2	400–500
	Nymphs	18–50	3–5	1–2	202
	Adults	16	–	–	–
Deilephila euphorbiae	Larvae	traces	–	–	–
Deilephila elpenor	Nymphs	–	–	–	800–1900
Galleria mellonella	Larvae	–	21	–	1700
Hyalophora cecropia	Larvae	–	–	–	1200
Hyalopgora cecropia	Nymphs	–	0–8	–	400–600
	Adults	–	–	–	650–1150
Telea polyphemus	Larvae	–	–	–	1306
Diptera					
Gasterophilus intestinalis	Larvae	95	10	184–294	–
Calliphora erythrocephala	Larvae	–	–	traces	–
Phormia regina	Larvae	–	70–125	–	absent
	Adults	–	up to 600	–	598

(Friedman[295]) and in the feeding larva of Celerio euphorbiae (Diptera) (Mochnacka and Petryszyn[296]).

On the other hand, if it is true that the utilization of trehalose as the circulating form of carbohydrates is widespread in insects, trehalose is synthesized in other invertebrates (Fairbairn[297]); however, as far as we know only in insects is trehalose inserted in a system of metabolic regulations characteristic of the phylum.

Occasionally, sugars other than trehalose have been found in the hemolymph of a number of insects. High concentrations of glucose have been found in the hemolymph of adults of *Apis mellifica* (600–3200 mg/100 ml) (von Czarnowsky[298]), in larvae of *Phormia regina* (larvae, 70–125 mg/100 ml; adults up to 600 mg/100 ml) (Evans and Dethier[292]) and in *Agria affinis* (in larvae 80 per cent of total carbohydrate appears as glucose) (Barlow and House[299]). Fructose is found at high concentration in the hemolymph of the adults of *Apis mellifica* (200–1600 mg/100 ml) (von Czarnowsky[298]) and in the larvae of *Gasterophilus intestinalis* (184–294 mg/100 ml) (Levenbook[300,301]).

The effect of diet on honeybees has been studied by Maurizio[302]. When the animals were fed on fructose or glucose the total blood-reducing sugar was either fructose or glucose.

Chang and Friedmann[303] have observed that when glucose or fructose were given in moderate amounts, trehalose was found in higher amounts than the reducing monosaccharide. When the supply of monosaccharide reached a high level, fructose or glucose exceeded trehalose, which was kept at the same level.

Trehalose, which appears to be used as the main form of carbohydrate transport in most insects, is synthesized in the fat body. The synthesis proceeds from G-6-P obtained from different sources in the gut and in the fat body through trehalosephosphate synthetase (Candy and Kilby[304]).

$$\text{G-6-P} + \text{UDP-glucose} \rightarrow \text{trehalose-6-P} + \text{UDP}$$
$$\text{trehalose-6-P} + H_2O \rightarrow \text{trehalose} + P$$

Maximum activity of trehalose-6-phosphatase occurs in the fat body of *Locusta migratoria* compared with other tissues (Mikolashek and Zebe[305]).

An inverse relationship exists between glycogen and trehalose concentration in the fat body, the former decreasing sharply at each molt or during starvation while the amount of trehalose in the fat body and in the hemolymph is kept constant (Duchâteau-Bosson *et al.*[306]; Saito[307]). The rate of trehalose synthesis and the supply of trehalose to the hemolymph as well as the trehalase activity of the blood are controlled by a hormone (hyperglycemic hormone) produced by the corpora allata (Steele[308]).

Murphy and Wyatt[309] have suggested that trehalosephosphate synthetase of *Hyalophora cecropia* is subject to allosteric regulation by G-6-P and by trehalose, the latter occupying an allosteric site on the enzyme and the former activating it.

In vertebrates, the cells have little glucose but this fermentable sugar is the form in which the blood carries carbohydrate for cellular use. Blood glucose is liberated by the liver cells from glycogen, owing to the presence of glucose-6-phosphatase in the cells. The glycogen itself is derived from a variety of substances: glucose, fructose, galactose, including some non-glucidic sources: amino acids, pyruvate, lactate, etc. (gluconeogenesis). This is due to the presence of fructose-1,6-diphosphatase, which allows for glucose-1-phosphate and glycogen biosynthesis by a reversal of glycolysis. In vertebrates, therefore, the circulating carbohydrate is glucose, and it is largely of endogenous origin, being the product of gluconeogenesis. From there the glucose goes to the cells. The whole system is controlled by a hormonal system including the specialized biosynthesis and activity of insulin, glucagon and adrenalin, through which the homeostasis of blood glucose concentration is insured, through a regulation of its liberation from liver glycogen and its cellular utilization.

Such a system is entirely foreign to insects. The regulation appears in their case to be focused at the level of trehalosemia. In *Carausius morosus*, the corpora cardiaca and the corpora allata control metabolism in the development until the penultimate stage. When the corpora allata and the corpora cardiaca are removed, trehalosemia decreases and the glycogen of tissues increases, while removal of the corpora allata alone leads to a decrease in protein synthesis (L'Hélias[310]). Steele[308] had noticed a hyperglycemia after introducing extracts of corpora cardiaca (due to activation of the fat body phosphorylase by the corpora cardiaca and the resulting synthesis of trehalose from glycogen) and McCarthy and Ralph[311] have observed a hyperglycemic activity of the corpora allata. This has been confirmed by Steele[312]. The hyperglycemic activity of extracts of corpora cardiaca in *Phormia* has been found by Stanley[313] to occur only when the insects are submitted to a 24-hour fasting period. The sugar content of the insect crop has also been found to be a limiting factor (Gelperin[314]).

In the adult *Calliphora erythrocephala* and *Aedes aegypti* the hyperglycemic activity is located at the level of the cells of the pars intercerebralis (Thomsen[315]). When these cells are removed, the glycogen accumulates in the fat body. It may be concluded that, depending on the species, the hyper-glycemic factor is localized either in the neurosecretory cells, in the corpora cardiaca or in the corpora allata, *i.e.* in a part of the retrocerebral factor.

Glucose injected into pupae is incorporated into glycogen whereas if they are treated with ecdysone, the glucose is incorporated into trehalose (Kobayashi and Kirimura[316]). The metabolism of carbohydrates in insects is subject to regulation at the levels of glycogen, of trehalose, and of chitin.

That the trehalose of the hemolymph may be derived from the glycogen of the fat body appears from a number of observations. At each period of intense activity or of fasting, when trehalosemia is lowered, the content of glycogen in the fat body also diminishes, and in much greater proportions (Duchâteau *et al.*[306]; Saito[307]). When, as shown by Steele[317], a rise in trehalosemia results from an injection of a corpora allata extract to *Periplaneta americana* there is a simultaneous fall in fat-body glycogen. But it appears that trehalose is not necessarily biosynthesized from G-6-P derived from glycogen. Bricteux-Grégoire *et al.*[318,319] have observed that in more or less undernourished silkworms, the radioactivity of labelled pyruvate, injected into the hemolymph, is found, four hours after the injection, only in the hemolymph trehalose and not in the fat-body trehalose. If labelled G-1-P is injected, it is also only at the level of hemolymph trehalose that the most intense incorporation is found. The glycogen of the fat body is likewise radioactive four hours after the injection of labelled G-1-P, but glycogen synthesis is twice as great in normally fed silkworms than in underfed ones. These results show that the amount of food consumed plays a role in the determination of the biosynthetic pathway followed by the materials involved in gluconeogenesis. This may be attributed to the control exerted by the level of trehalosemia on the synthesis of trehalose and of glycogen. As stated above, above certain concentration limits, trehalose inhibits trehalosephosphate synthetase so that the resulting increase in UDPG concentration would activate glycogen synthesis (Murphy and Wyatt[320]).

That trehalose is the circulating carbohydrate (which is utilized and subject to regulation) in insects is confirmed by a number of observations:

(*a*) the cells of many tissues of insects contain an active trehalase: muscles, gut, salivary glands (Kalf and Rieder[321]; Howden and Kilby[322]; Zebe and McShan[323]; Friedman[324]; Duchâteau-Bosson *et al.*[306]; Gussin and Wyatt[325]; Gilby *et al.*[326]; Stevenson[327]; Hansen[328]).

(*b*) in the course of muscular activity (sustained flight, for example) the concentration of plasma trehalose is lowered: the muscles therefore utilize this trehalose (Evans and Dethier[292]; Clegg and Evans[329]; Bücher and Klinkenberg[330]).

(c) in the nymphs of *Deilephila elpenor*, just after the termination of diapause, when the adult tissues start growing, the trehalosemia diminishes (Duchâteau and Florkin[331]).

The cells of the epidermis seem to lack trehalase (Zebe and McShan[323]; Duchâteau-Bosson et al.[306]). The cyclic activity of the epidermal cells of insects (molting) is paralleled by the variations in hemolymph trehalase, which remains inhibited during the whole intermolt period (Friedman[324]). The inhibition of the hemolymph trehalase is suppressed at the beginning of every molting period and a drop in trehalose concentration occurs in the hemolymph (Howden and Kilby[322]; Duchâteau-Bosson et al.[306]). The

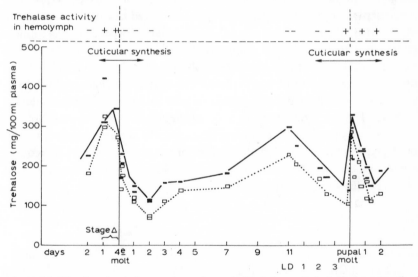

Fig. 32. Modification of trehalose concentration and of trehalase activity in the hemolymph of *Bombyx mori*. ■, anthrone reactive material, in mg trehalose/ml; □, trehalose, mg/100 ml; L.D., last defecation. (Duchâteau-Bosson et al.[306]).

glucose concentration rises in the hemolymph and is used not only for metabolic purposes but also for the synthesis of chitin in the epidermis (Candy and Kilby[304, 332]). The variation of trehalase activity and of trehalose concentration in the hemolymph of *Bombyx mori* during the 4th larval molt and the pupal molt is represented in Fig. 32 (Duchâteau-Bosson et al.[306]).

In *Samia cynthia ricini* larvae the inhibited trehalase of the hemolymph has been observed to be released at the time of molt (Chang et al.[333]). In the adult of *Phormia regina*, Friedman[324] has shown that the trehalase is maintained in the inactive state by a proteinaceous inhibitor.

20. Sustained flight in bees and flies

Carbohydrate is the only fuel used in the very active phenomenon of sustained flight to be observed in some insects (Hymenoptera and Diptera). It is supplied to the insect muscles in the form of non-reducing trehalose which represents a blood reserve used in the great, instantaneous increase of respiration that occurs when the insect attains full flight. This adaptation is the result of the acquisition by the species concerned of a lateral extension on the glycolytic pathway. While in the general pattern of this pathway, NADH reduces pyruvate, in the species in question it reduces dihydroxyacetone phosphate in the presence of a NAD-dependent cytoplasmic 3-glycerophosphate dehydrogenase. The 3-glycerophosphate produced diffuses into mitochondria where the 3-glycerophosphate dehydrogenase is located and directly enters the respiratory chain. The system is characterized by a relative lack of lactate dehydrogenase. This should lead to an accumulation of NADH and consequently to a deficit in the NAD^+ necessary for the oxidation of glyceraldehyde-3-phosphate. But the increased concentration of 3-glycerophosphate dehydrogenase prevents this effect. It reduces dihydroxyacetone phosphate to 3-glycerophosphate while oxidizing NADH to NAD^+ (literature, see Hansford and Sacktor[334]).

21. Glycerol production in diapausing insects

The accumulation of 3-glycerophosphate just described in the case of the shunt, of which advantage is taken in rapid flight, is likewise taken advantage of in the diapause of insect nymphs (Schneiderman and Williams[335]) or of eggs. Andrewartha[336] has reviewed the biological aspects of the adaptation called diapause. In insects adapted to the climate of the cool temperate zone of the Northern Hemisphere, diapause occurs in the stage of meta-morphosis coinciding with the lowest temperatures in the succession of seasonal changes. During the nymphal diapause undergone by a number of Lepidoptera, the oxygen intake falls to 1/50 of the prediapausal level and is no longer modified by carbon monoxide, cyanide or azide, all of these being inhibitors of cytochrome oxidase. This results from a lowering of the concentration of cytochrome c, a consequence of which is a large excess of cytochrome oxidase relative to cytochrome c. In this situation, the low level of cytochrome c causes a fresh formation of NADH, which is reoxidized in the presence of a 3-glycerophosphate dehydrogenase; the resulting 3-glycerophosphate is dephosphorylated into glycerol in the tissues.

This glycerol protects the tissues of the insect from the effects of a lowered temperature. In the sawfly *Bracon cephi* the concentration of glycerol in the tissues reaches values as high as 25 per cent of fresh tissues, which allows the animal to withstand such low temperatures as $-40°C$ (Salt[337]).

In the nymphal diapause of *Hyalophora (Samia) cecropia*, the concentration of glycerol in the hemolymph may reach 3.5 per cent. This glycerol is derived from glycogen *via* trehalose. It appears in the hemolymph on the day of nymphosis and increases until it reaches a maximum after one or two months[338,339].

In another saturniid, *Philosamia (Samia) cynthia*, the diapausing nymphs accumulate less glycerol, but if exposed to lower temperatures the glycerol increases in the hemolymph[340].

Glycerol concentration is not the only biochemical change observed in hibernating nymphs. Data on free amino acids in larvae and nymphs in diapause are available for *Sphinx ligustri* and *Smerinthus ocellatus*[237]. Alanine levels are much higher in these diapausing nymphs than in the corresponding larvae, another aspect of the protection against low temperature effects. This is not the case with non-hibernating nymphs of *Euproctis chrysorrhea* nor, except for a short time following nymphosis, for the non-hibernating nymph of *Bombyx mori*. Other cases of freezing-point depressors are known such as the role of small glycoproteins as antifreeze in the blood of antarctic fishes[341]. There are also cases of other physiological radiations of glycerol. In *Artemia salina* the process of kyst rupture and the liberation of the embryo results from an increase in glycerol concentration[342].

22. Heat generation in bumblebees (*Bombus vagans*)

Unlike bees belonging to the genus *Apis*, bumblebees are able to forage on cold days. This is the result of their power (which they owe to a particular feature of their glycolysis pattern) to keep their muscles warm[343]. As was shown by Newsholme *et al.*[344], the flight musculature of *Bombus* contains very high amounts of fructose-diphosphatase (up to 40 times higher than in vertebrate muscles). As soon as the F-1,6-PP is generated

$$F\text{-}6\text{-}P + ATP \rightleftharpoons F\text{-}1,6\text{-}PP + ADP$$

the molecule of F-1,6-PP is split into F-6-P and P_i, the net reaction being:

$$ATP \rightleftharpoons ADP + P_i + \text{heat}$$

In general, in glycolysis, fructose-diphosphatase and phosphofructokinase are not operating simultaneously as fructose-diphosphatase is usually inhibited by AMP and since phosphofructokinase is activated at low-energy charges of the adenylate system. The bumblebee possesses a high concentration of AMP-insensitive fructose diphosphatase.

23. 2,3-Diphosphoglycerate formation in mammalian erythrocytes

This substance modifies the oxygen capacity of hemoglobin and is formed through a modified glycolytic pathway described by Brewer and Eaton[345]. As the concentration of 2,3-diphosphoglycerate rises, the oxygen capacity falls, due to a combination of one mole of 2,3-diphosphoglycerate with hemoglobin (a ligand-induced configuration change of the tetramer). The concentration of the modulator appears higher in venous blood and its oxygen capacity lower, a factor which favours oxygen delivery to tissues.

The transformation of 1,3-diphosphoglycerate into the 2,3-compound relies on the presence of the enzyme diphosphoglycerate mutase.

The negative modulation acting on oxygen capacity in venous conditions appears to be accounted for by inositol hexaphosphate in birds and reptiles and by ATP in fish. In vertebrates, 2,3-diphosphoglycerate, inositol hexaphosphate and ATP act, in the aspect considered, analogously.

24. Pathway from glucose to fructose in mammalian foetus and spermatozoa

As Hers[346] has shown, glucose is transformed in this pathway into sorbitol (coenzyme: NADP) and sorbitol into fructose (coenzyme: NAD).

In the placenta of some mammalian species, e.g. sheep and cow, glucose is converted to sorbitol which is carried to the liver of the foetus where it is converted to fructose.

In the seminal vesicles of mammals fructose is also derived from glucose.

C. THE EVOLUTION OF THE INTRACELLULAR PATHWAY OF HYDROGEN TRANSPORT

25. Introduction

NADH resulting from glycolysis or of dehydrogenations in the Krebs cycle can be used in a number of different ways, either in the same cell or in different categories of cells or organisms. One important feature of the central metabolic pattern is the entrance of electrons (carried on hydrogen atoms) from intermediates of the Krebs cycle into the respiratory chain. Other metabolic reducing effects of NADH are well-documented such as the reduction of pyruvic acid, of acetaldehyde, of phosphoglyceraldehyde, of phosphodihydroxyacetone, etc.

Another use of NAD is the part it plays in the pentosephosphate cycle which produces NADPH, used in reductive biosyntheses. Besides these multiple roles in the framework of the general biochemical pattern of cells. a number of other multiple uses of NAD have developed in the course of animal evolution.

26. Provisions for exploiting anoxic environments

When oxygen is lacking, vertebrate striated muscles resort to the Pasteur effect (increase in anaerobic capacity of ATP formation when oxygen is lacking).

This procedure has been greatly perfected in the diving of turtles, for instance. In this phenomenon three phases are distinguished, of which the third is totally anoxic (Jackson[347]) and relies on compensatory anaerobic ATP formation.

In obligate anaerobic forms, such as *Ascaris*, the obligate anaerobic pathway of ATP formation has been favoured by a deletion of pyruvate kinase and lactate dehydrogenase (see Chapter I).

27. Cold adaptation in homeotherms

In homeotherms, one aspect of the regulation of the internal medium (homeostasis) is the maintenance of its temperature with close limits. This regulation presents a number of physiological aspects and also a molecular aspect in which the utilization of NADH in the respiratory chain is modified.

References p. 220

Maximum efficiency of the respiratory chain is insured when the ratio P:O is equal to 3, *i.e.* when 3 molecules of inorganic phosphate are utilized for the synthesis of three ATP (from ADP) in the transfer of a pair of electrons to an atom of oxygen. In these conditions, the free energy produced by the reduction of oxygen ($\pm$ 52 kcal·M^{-1}) is recovered in the form of ATP to the extent of approximately 40 per cent, the rest being dissipated as heat. There are three sites of phosphorylation along the respiratory chain and each one may be shunted if, for instance, cytochrome *c* accepts electrons from NADPH and if consequently the P:O is reduced from 3 to 1. The energy that could have been utilized to generate ATP is partly (2/3) added to the amount of dissipated heat (literature, see Smith[348]). Under the stress of cold, a homeotherm animal shows an increase not only in oxygen consumption but also in thyroxine production. Not only does thyroxine uncouple the oxidative phosphorylation, it also inhibits the transhydrogenase responsible for the reduction of NAD by NADPH. The molecular aspect of cold adaptation in homeotherms therefore results from the existence of two different pathways of hydrogen transport within the cells of homeotherms. In thyroid-stimulated thermogenesis, the source of dissipated heat is ATP split by the Na^+/K^+ ATPase of the cell membranes of liver, muscle and kidney (Ismail-Beigi and Edelman[349]), which is activated by the thyroid hormones. This is an instance of physiological radiation of the ATPase system.

28. Brown adipose tissue (brown fat)

Brown adipose tissue which is found in mammals surrounds such vital organs as the heart and contains large numbers of circular mitochondria with elaborated cristae.

In brown adipose tissue the oxidative phosphorylation is uncoupled and the free energy derived from fat in the respiratory chain, instead of being stored in ATP, is dissipated as heat. It appears that epinephrine regulates the furnace function of brown adipose tissue[350], useful in early post-natal life, in cold acclimatization as well as in arousal from hibernation.

29. Loss of respiratory chain in trypanosomes of the *brucei* group

Trypanosomes of this group have lost the respiratory chain, which is replaced by a different system. Grant and Sargent[351] found that it is possible to separate water lysates of *T. rhodesiense* into a particulate fraction and a particle-free fraction. In the particulate fraction they found an enzyme oxidizing L-α-glycerophosphate at the expense of oxygen without participation of the usual respiration chain. This oxidizing system, which does not produce H_2O_2, involves FAD and iron (Bide and Grant[352]). In the presence of catalytic amounts of L-α-glycerophosphate or dihydroxyacetone phosphate the action of α-glycerophosphate dehydrogenase and of α-glycerophosphate oxidase maintains an oxidation–reduction cycle.

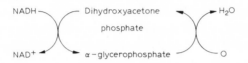

V. SOME EXTENSIONS ON THE TERPENOID AND STEROID PATHWAYS OF ANIMALS

A. INTRODUCTION

In Chapter I (Vol. 29 A), we have described several catenary biosyntagms, among them the terpenoid and steroid metabolic biosyntagms, represented in Fig. 25 (Vol. 29 A, p. 83). In the same chapter we described the terminal extension leading, in animals, to bile salts and to mammalian steroid hormones, and we refer the reader to the relevant sections.

The synthesis of sterols through the terpenoid biosyntagm is a general feature of eucaryotes and its significance is suggested by Bloch[353] as related to their association with the composition of the cytoplasmic membranes of these organisms. There are similar structural requirements in all cases (a planar ring system basically of the cholestane, ergostane or stigmastane type, a 3 β-hydroxyl group). The general presence (except in case of secondary loss or inhibition of genetic powers) of the metabolic terpenoid–steroid pathway leading to the biosynthesis of an essential constituent of the intracellular membranes of eucaryotic cells, cholesterol, leads to a variety of extensions on the catenary pathway.

As already pointed out by Bloch[354] in 1964 and illustrated in Fig. 27 (Vol. 29 A, p. 86), the catenary biosyntagm subserving the mevalonic pathway reaches the step of lanosterol in bacteria as well as in mycetes, plants and animals. From this step on, all eucaryotes synthesize sterols as obligatory constituents of their cytoplasmic membranes.

The characteristic biosyntagm of animals show terminal and lateral extensions on the biosynthetic assembly line. The reader is referred to Chapter I (Vol. 29 A) for a description of the terminal extensions leading to bile salts and to mammalian steroid hormones.

B. LOSS OF THE CAPACITY TO SYNTHESIZE STEROLS

While vertebrates (with the possible exception of elasmobranchs[354]) appear to be able to biosynthesize sterols, this capacity is lacking in insects. Insects do not synthesize sterols but they do require them in their food. This problem is solved by the compensatory action of an enzymatic system able to transform phytosterols in the food into cholestane derivatives that fit the intracellular membrane fabric.

This system which has been identified in many insects removes the C-24 alkyl substituents on the phytosterol side-chain[356]. Certain insects have acquired the enzymes necessary for the saturation of double bonds or their introduction into sterol nucleus.

C. TERMINAL EXTENSIONS FROM LANOSTEROL

30. Steroid neurotoxins of Echinoderms

The neurotoxins of the sea cucumbers (Holothuridae) are believed to be metabolic derivatives of lanosterol. One of these neurotoxins is 22,25-oxidoholothurinogenin (Chanley et al.[357]).

22,25-Oxidoholo-
thurinogenin

31. Steroid neurotoxins of amphibians

These are very poisonous substances that function as repellents. They have been studied mainly in salamanders, from which a series of related molecules

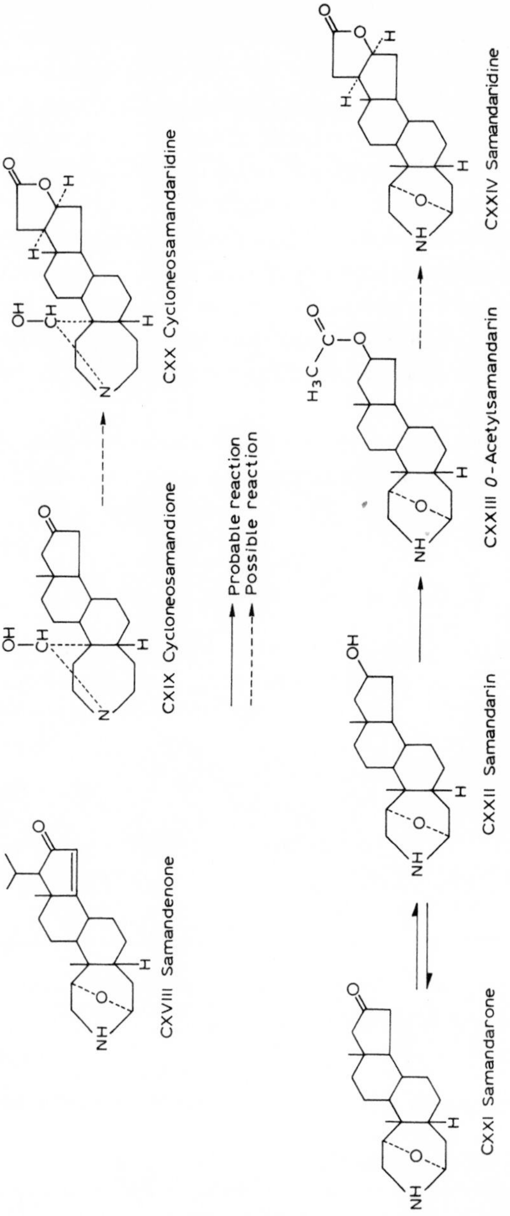

Fig. 33. Amphibian steroid alkaloids. (Balinsky[180])

have also been isolated. The structure of samandarin has been determined by Habermehl[358]. Samandarin is biosynthesized from cholesterol as was demonstrated by using radio-active labelled cholesterol (Habermehl and Hauf[359]). The heterocyclic ring results from a fission of ring A of cholesterol between carbons 2 and 3, of an insertion of nitrogen coming from glutamine, and of a closure of the ether bridge from the C-3 hydroxy group to C-1.

The 16-dehydro derivative of samandarin (CXII in Fig. 33) samandarone (CXXI in Fig. 33) (Wölfel *et al.*[360]) has also been found in salamanders, as well as samandaridine (LXXIV) (Habermehl[361]) and samandenone (CXVIII) (Habermehl[362]) which contain 21 and 22 carbon atoms respectively. Steroid alkaloids are not restricted to either salamanders or to urodeles. Samandarin has been found in the frog *Pseudophryne corrohoree* (Habermehl[363]).

Batrachotoxin, the poisonous protective alkaloid of frog species of the *Phyllobates* genus, used by Colombian Indians as arrow poison is also a steroid alkaloid.

32. Cardiac-active steroids in toad skin

Toads in their defence mechanisms use the highly toxic bufogenins, such as bufotalin, which is biosynthesized from cholesterol (Siperstein *et al.*[364]).

Bufotalin

Bufotalin, in the form of its 14β-suberylarginine conjugate, is the major genin of the bufotoxin of *Bufo vulgaris*.

33. Steroids produced as defensive secretions by water beetles

The aquatic beetles of the family Dysticidae discharge a fluid containing steroids produced by a pair of glands in the neck region and toxic to fish and Amphibia. This fluid has been found to contain C_{21} corticosteroids in several cases (Schildknecht and Hotz[365]; Schildknecht *et al.*[366]) and, in one case,

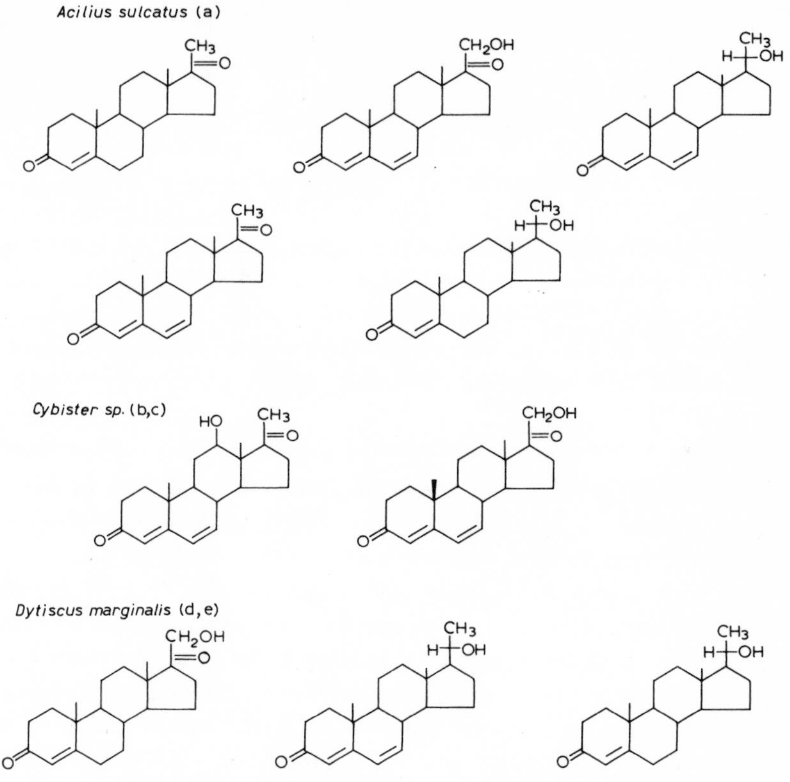

Fig. 34. Steroid defensive secretions of water beetles. (Clayton[368])

the C_{17} androstene, testosterone (Schildknecht et al.[367]). A series of other water beetles have been found to liberate steroid defensive secretions (Fig. 34).

34. Insect and crustacean ecdysones

Besides their role in the structure of the intracellular membranes, sterols in Crustacea and insects are the precursors of the hormone ecdysone which promotes molting and adult development. Ecdysones are C_{27} or C_{29} compounds. The natural animal ecdysones so far isolated possess (with the sole exception of 2-deoxy-β-ecdysone) a tetracyclic nucleus bearing hydroxyl groups in positions 2 β, 3 β and 14 α, an unsaturated (Δ^7)-6-ketone system and cis fusion between rings A and B.

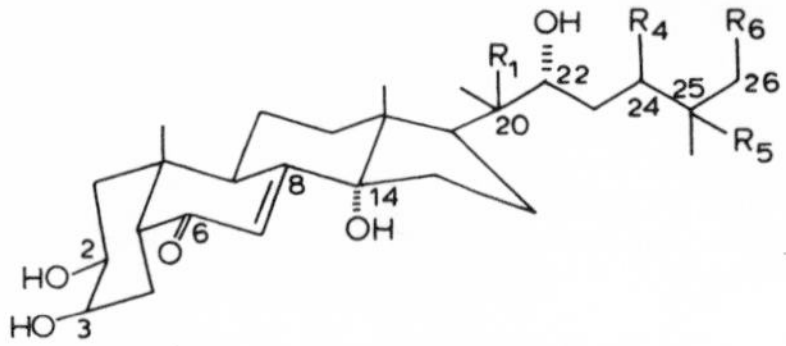

		R_1	R_4	R_5	R_6
I	α-Ecdysone			OH	
II	β-Ecdysone	OH		OH	
III	2-Deoxy-β	OH		OH	
IV	26-Hydroxy-β	OH		OH	OH
V	Callinecdysone-A	OH			OH
VI	Callinecdysone-B	OH	Me	OH	

It is generally believed that the zooecdysones are secreted by insect prothoracic glands or crustacean Y-organs but it would be premature, as Bern[369] emphasizes, to think that they alone are responsible for it.

It has been shown by Karlson and Hoffmeister[370] that, in mature *Calliphora* larvae, cholesterol is the precursor of α-ecdysone. Karlson[371] has suggested that cholesterol may be a starting material for ecdysone bio-synthesis, possibly *via* 7-dehydrocholesterol resulting from a dehydrogenation. The last step in ecdysone biosynthesis is the hydroxylation of α-ecdysone at C-20 (King and Siddall[372]; Thomson et al.[373]). The conversion of α-ecdysone into β-ecdysone (ecdysterone) as shown in Fig. 35 by actively molting *Crangon nigricanda* (a shrimp), by premolt fiddler crab *Uca pugilator*

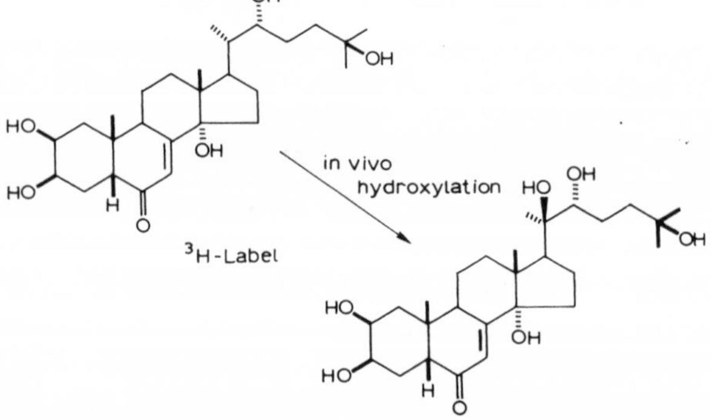

Fig. 35. Conversion of α-ecdysone to β-ecdysone. (Siddall[374])

and by fifth instar of the blowfly *Calliphora vicina* represents the last step in ecdysterone biosynthesis in crustaceans and insects.

D. LATERAL EXTENSIONS

35. Insect juvenile hormone

The endocrine system of insects is basically composed of the brain, the prothoracic glands and the corpora allata. While, as stated above, the prothoracic glands secrete ecdysone which induces molting, the corpora allata secrete the juvenile hormone which appears to dictate the suppression of the information carried by messenger RNA for adult cuticular protein. Juvenile hormone appears to influence the replication of DNA, which determines form after ecdysis. The hormone has been isolated in two forms a C_{16} (II) and a C_{17} (I). It is an isoprenoid of the following structure (very similar to methylfarnesoate epoxide).

The natural juvenile hormones.
(I) R = Et   Röller *et al.* (1967)
(II) R = Me   Meyer *et al.* (1968)

It appears that juvenile hormone is metabolically derived from a methylation, by δ-adenosyl-methionine, of sesquiterpene, which would account for the existence of the C_{16} and the C_{17} forms though no definite evidence for it has as yet been produced.

36. Terpenoid coactones used by insects

Besides a number of sex pheromones, releasing sexual attraction or sexual behaviour or both (see Wilson and Bossert[375]), various terpenoids are used by insects as coactones in communication and defence (for literature, see Wilson and Bossert[375]; Weatherston[376]; Regnier and Law[377]; Florkin and Schoffeniels[378]).

Fig. 36 shows some examples of terpenoid alarm and defence substances of insects; Fig. 37, some terpenoid attractants in insects.

The alarm and defence substances include the ant alarm substances

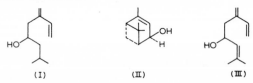

Fig. 36. Some terpenoid alarm and defence substances of insects. Ants. I. Citronellal; II. Citral *(Acanthomyops claviger)*; III. Limonene *(Myrmicaria natalensis)*; IV. Iridomyrmecin; V. Isoiridomyrmecin; VI. Iridodial; VII. Dolichodial (members of Dolichoderinae); VIII. Dendrolasin *(Lasius fuliginosus)*. Beetles: IX. Cantharidin (Meloidae). Bees: X. Isoamyl acetate *(Apis mellifera)*. Termites: XI. α-Pinene *(Nasutitermes sp.)*. (Clayton[368])

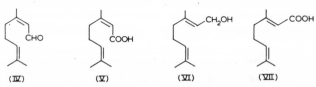

Fig. 37. Some terpenoid attractants of insects. Beetles: I. 2-methyl-6-methylene-7-octene-4-ol; II. *cis*-verbenol; III. 2-methyl-6-methylene-2,7-octadien-4-ol *(Ips confusus)*. Bees: IV. citral; V. nerolic acid; VI. geraniol; VII. geranic acid (food attractants of *Apis mellifera*). (Clayton[368])

(citronellal, citral, limonene, iridomyrmecin, isoiridomyrmecin, iridodial) the cantharidin of meloe beetles, the dolichodial (or anisomorphal) of *Anisomorpha*, the isoamylacetate released by a stinging bee and which incites other bees to sting, and the α-pinene of termites.

All these terpenes are believed to be derived by lateral extensions on the terpenoid and steroid pathway. The incorporation of mevalonic acid into citronellal and citral of *Acanthomyops* has been reported[379], as well as into dolichodial of *Anisomorpha*[380]. 6-Methyl-5-hepten-2-one[381], limonene[382] and 4-methyl-2-hexanone[383] all arise by the normal terpene pathway.

In the frass produced by the male of *Ips confusus* boring in ponderosa pine *(Pinus ponderosa)*, there is a powerful assembling scent evoking a mass attack by individuals of both sexes. The females are, however, more responsive than the males. When the insects arrive at the source of attraction, they take part in boring, feeding, mating and oviposition (Wood *et al.*[384]). Three terpene alcohols have been identified and synthesized (Silverstein *et al.*[385]) (see Fig. 37). They are inactive individually but produce a response if two of them are used. In the volatile scent of the male bumblebees of the species *Bombus terrestris* the main component is 2,3-dihydro-6-*trans*-farnesol[386]. By this scent, the males perfume-mark their flight territory and gather the females.

$$H_3C \atop H_3C \quad \diagup C{=}CH{-}CH_2{-}CH_2{-}\underset{\underset{CH_3}{|}}{C}{=}CH{-}CH_2{-}CH_2{-}\underset{\underset{CH_3}{|}}{CH}{-}CH_2{-}CH_2OH$$

E. PHYSIOLOGICAL RADIATIONS

37. A physiological radiation of ecdysone

Cleveland[387], employing crystallized ecdysone, reported the induction by this hormone of gametogenesis marking the onset of sexual phenomena in parasitic *Trichonympha*, the intestinal flagellate of the xylophagous roach *Cryptocercus*.

38. Physiological radiations of carotenoids and their derivatives

Plants are able to synthesize tetraterpenes and carotenoids while animals are able only to modify them, *e.g.* by oxidation. In plants the carotenoids are synthesized in the terpenoid and steroid pathway. This lateral extension, grafted on the pathway at the level of isopentylpyrophosphate and leading to carotenoids (C_{40}) *via* geranyl-geraniolpyrophosphate (C_{20}) is lacking in animal cells. A cell capable of photosynthesis contains at least one chlorophyll and at least one yellow pigment. The pathways of carotenoid biosynthesis in plants may be essentially regarded as directed towards the synthesis of accessory photosynthetic pigments. Advantage is taken in photosynthesis and in the phototropism of plants of the fact that carotenoids can absorb light and transfer energy. In animals, other properties of the carotenoids and of their derivatives are utilized in a number of different physiological radiations which do not involve any structural changes. The animals make use in vision of their capacity to undergo *cis–trans* isomerization.

The visual pigments are chromoproteins that bleach rapidly when exposed to light. This process is normally counterbalanced by regeneration in the living eye.

As was mentioned in section 30 (b) of Chapter I (Vol. 29 A, p. 82), there are two main series of naturally occurring visual pigments: rhodopsins and porphyropsins. The structure of their chromophorically active moiety has been clarified by showing that, under the action of light, retinol (vitamin A) is released from rhodopsin and 3-dehydroretinol (vitamin A_2) from porphyropsin. The extraction of bleached retinas or the destruction of the visual pigment in darkness with chloroform yields another substance, retinaldehyde (retinene-1) in the case of rhodopsin and 3-dehydroretinaldehyde (retinene-2) in the case of porphyropsin.

The reactions, observed by Wald[388], are summarized in Fig. 38.

Like the carotenoids, retinol and its aldehyde may exist in a number of different configurations due to the numerous double bonds in the polyene chain (*cis–trans* isomerism).

There exist many rhodopsins and porphyropsins distinguished by the spectral locations of their α-bonds (Bridges[389]; Dartnall[390]; Crescitelli[391]; Morton and Pitt[392]). For rhodopsins the absorption maxima range from 433 and 440 nm in honeybee and frog (Goldsmith[393,394]; Denton and Wyllie[395]; Donner and Reuter[396]; Bridges[397]), to 562 nm in chicken (Wald *et al.*[398]), while that of porphyropsins occur over a more restricted range, from 510 nm in some labrid fishes (see Wald[399]) to 543 nm in the fish *Osmerus* (Bridges[389]).

The wide variety of pigments found in fishes has to do with their adaptation

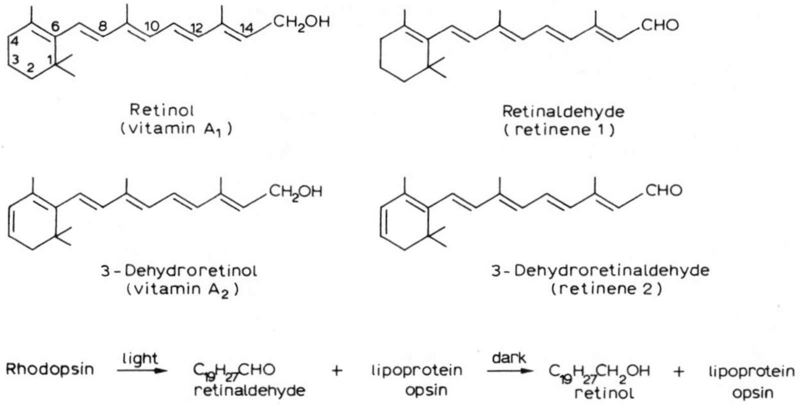

Retinol
(vitamin A$_1$)

Retinaldehyde
(retinene 1)

3-Dehydroretinol
(vitamin A$_2$)

3-Dehydroretinaldehyde
(retinene 2)

Rhodopsin $\xrightarrow{\text{light}}$ C$_{19}$H$_{27}$CHO + lipoprotein $\xrightarrow{\text{dark}}$ C$_{19}$H$_{27}$CH$_2$OH + lipoprotein
retinaldehyde   opsin   retinol   opsin

Fig. 38. Action of light on rhodopsin. (Wald[388])

to the diversity of spectral light distributions in their aquatic habitats. Fishes living in deep marine water where solar radiation is limited to a narrow band in the bluegreen at 480 nm have rhodopsins absorbing in the same spectral region. Fresh-water fishes live in a medium where there is a high proportion of red light, and their pigments have absorption peaks displaced toward the red. Fishes living in intermediate habitats as well as land vertebrates usually have pigments absorbing near 500 nm corresponding to the wavelength of the solar radiation maximum at the earth's surface. Porphyropsins are generally found in fresh-water forms, in accordance with the nature of their photic environment. They may be the only visual pigment as is the case for *Cyprinus* (Crescitelli and Dartnall[400]), for *Lepisosteus* (Bridges[401]) and for *Necturus* (Crescitelli[402]). But many fresh-water fishes have a mixture of visual pigments and an interconversion between rhodopsin and porphyropsin may be observed in answer to changes of light intensity or to other modifications of environmental factors. Temperate-zone fresh-water fishes, having a mixture of visual pigments, increase the proportion of rhodopsin in summer. This has been demonstrated in the Cyprinid fish *Notemigonus crysoleucas bascii* (Bridges[403]). During the change of different light habitats in the metamorphosis of amphibians from aquatic tadpole to terrestrial adult, a porphyropsin–rhodopsin interconversion also takes place (Wald[388]). Wilt[404] has proposed a direct role of thyroxine in this phenomenon. Naito and Wilt[405], using tritiated retinol, have shown that it can be a precursor of 3-dehydroretinaldehyde. By injecting the labelled compound in the isolated surviving eye of the sunfish, *Lepomis* sp., a fresh-water species characterized by a mixture of visual pigments in which porphyropsin predominates, they were able to locate radioactive 3-dehydroretinaldehyde. They also showed that the introduction of thyroid hormone in the eye reduces the rate of transformation.

Whether the conversion takes place between retinol and 3-dehydro-retinol or between retinaldehyde and 3-dehydroretinaldehyde is not clear. However, the liver is unable, under the conditions of organ culture, to convert retinol to 3-dehydroretinol. These experiments combined with the results obtained by Wilt suggest that the bullfrog tadpole converts retinol to 3-dehydroretinaldehyde in the eye with a formation of porphyropsin. The thyroid hormone would reduce or abolish this formation and favour the utilization of retinaldehyde to form rhodopsin.

Lampreys, eels and salmon migrating between fresh- and sea-water also exhibit porphyropsin–rhodopsin interconversions. Salmon and trout (different species of the genera *Oncorrhynchus* and *Salmo*) have mixtures of retinaldehyde and 3-dehydroretinaldehyde (Wald[406]; Kampa[407]; Bridges[408]; Munz and Beatty[409]). Beatty[410] has shown that in salmon the two visual pigments are not always present in the same proportion and that the predominant one is not always porphyropsin. He observed that in the spawning migration of adult salmon there was a progressive increase in the percentage of porphyropsin absorbing at 527 nm in the retina, and this increase (except for the sockeye salmon) resulted in a conversion from a retina having predominantly a rhodopsin absorbing at 503 nm to one with a preponderance of porphyropsin (526 nm).

Some relationship between photic environment and spectral sensitivity of visual pigments may also be found in Decapoda. In *Homarus vulgaris*, the spectral sensitivity of the eye is greatest between 516 and 531 nm, a range "very similar to the spectral composition of attenuated daylight in the water and depths where lobsters live" (Kampa *et al.*[411]). The eye of the hermit crab *Eupagurus bernhardus* has its maximum sensitivity near 500 nm (Stieve[412]), a property well adapted to the photic environment provided in shallow coastal waters.

In conclusion, it may be stated, in the words of Munz[413] as follows:

"Several investigators have concluded that the visual pigments of deep-sea animals (teleosts, crustaceans and elasmobranchs) have become adapted to the predominantly blue light in that habitat by shifts in λ_{max} to wavelengths below 500 nm. This type of evolutionary response to the photic environment has occurred in other marine habitats. Some pelagic fishes living near the surface also have so-called "deep-sea rhodopsins" well suited to the predominantly blue light in this environment. In turbid, greenish, or yellowish coastal waters, fishes have "rhodopsins" with λ_{max} above 500 nm. These ecological groupings of fishes, with visual pigments correlated to the spectral distribution of sunlight in each habitat, also appear to be paralleled by the crustaceans.

In fresh water, both fishes and crustaceans have visual pigments adapted to the predominance of long wavelengths in the photic environment. In fishes, this is accomplished by the retinene-2 pigments, often mixed with the corresponding retinene-1 pigment. Mixtures of rhodopsin and porphyropsin, in changeable proportions may be adapted to the highly variable nature of the photic environment in many freshwater habitats."

Such broad generalizations are nevertheless not without exceptions as attested, for instance by the existence of rhodopsins with absorption peaks above 520 nm in terrestrial geckos (Crescitelli[414]). Furthermore, if the relation between the spectral properties of the visual pigments of vertebrates and of invertebrates is clear in many cases, it is not always so and we must await greater knowledge of the cone pigments, so far little known in vertebrates, in order to establish a complete correlation between pigment distribution and environmental factors.

ACKNOWLEDGEMENTS

For permission to reproduce copyright material, the author expresses his appreciation to the following: J. Balinsky; W. R. Butt and J. F. Kennedy; R. B. Clayton; M. O. Dayhoff; H. M. Grey; L. E. Hood, W. R. Gray, B. G. Sanders and W. J. Dreyer; A. E. Needham; Academic Press, New York; Cold Spring Harbor Laboratory; National Biochemical Research Foundation, Georgetown University Medical Center, Washington, D.C.

References p. 220

REFERENCES

1 M. Florkin and B. T. Scheer (Eds.), *Chemical Zoology*, Vols. 1–9, Academic Press, New York, 1967–1974.
2 J. Price, *The Origin and Evolution of Life*, English Univ. Press, London, 1971.
3 G. R. De Beer, in: *Evolution as a Process*, London 1954.
4 R. B. Clark, *Dynamics in Metazoan Evolution*, Oxford Univ. Press, 1964.
5 O. Tiegs and S. M. Manton, *Biol. Rev. Cambridge Phil. Soc.*, 33 (1958) 255.
6 W. Garstang, *Quart. J. Microscop. Sci.*, 72 (1928) 51.
7 G. S. Carter, *Syst. Zool.*, 6 (1957) 187.
8 A. S. Romer, *Quart. Rev. Biol.*, 21 (1946) 33.
9 P. P. Grassé (Ed.), *Traité de Zoologie*, Masson, Paris, 1948–1973.
10 M. O. Dayhoff (Ed.), *Atlas of Protein Sequence and Structure 1972*, Washington D.C., Nat. Biom. Res. Foundation, 1972, Vol. 5; Suppl. I, 1973
11 M. Calvin, *Trans. Leicester Lit. Philos. Soc.*, 62 (1968) 45.
12 M. Florkin, *A Molecular Approach to Phylogeny*, Elsevier, Amsterdam, 1966.
13 M. Florkin, in: G. Eglington and M. T. J. Murphy (Eds.), *Organic Geochemistry*, Springer, Berlin, 1969.
14 M. Florkin, in: R. Buvet and C. Ponnamperuma (Eds.), *Chemical Evolution and the Origin of Life*, North Holland, Amsterdam 1971.
15 P. H. Abelson, *Carcegie Inst. Wash., Yearbook*, 53 (1954) 97.
16 P. H. Abelson, *Carnegie Inst. Wash., Yearbook*, 54 (1955) 107.
17 P. H. Abelson, *Sci. Am.*, 195 (1956) 83.
18 P. H. Abelson, *Ann. N.Y. Acad. Sci.*, 69 (1957) 276.
19 H. C. Ezra and S. F. Cook, *Science*, 126 (1957) 80.
20 S. Ijiri and T. Fujiwara, *Proc. Japan Acad.*, 34 (1958) 280.
21 T. V. Drozdova and A. V. Kochenov, *Geokhimiya*, (1960) 748.
22 T. V. Drozdova and A. M. Blokh, *Geochemistry*, 3 (1966) 530.
23 W. G. Armstrong and L. B. H. Tarlo, *Nature*, 210 (1966) 481.
24 A. Oekonomidis, *Fol. Biochem. Biol. Graeca*, 5 (1968) 6.
25 T. Fujiwara, *J. Geol. Soc. Japan*, 67 (1961) 97.
26 M. Akiyama and T. Fujiwara, *Misc. Rept. Res. Inst. Nat. Resources, (Tokyo)*, 67 (1966) 67.
27 S. Hotta, *J. Geol. Soc. Japan*, 71 (1965) 842.
29 P. E. Hare and R. M. Mitterer, *Carnegie Inst. Wash., Yearbook*, 65 (1966) 362.
29 G. Szoor, *Acta Biol. Debrecina*, 5 (1967) 111.
30 S. M. Manskaya and T. V. Drozdova, *Geochemistry*, (1962) 1077.
31 D. B. Carlisle, *Biochem. J.*, 90 (1964) 1c-2c.
32 R. Fikentscher, *Zool. Anz.*, 103 (1933) 289.
33 M. Blumer, *Mikrochemie*, 36 (1951) 1048.
34 M. Blumer, *Nature*, 188 (1960) 11.
35 M. Blumer, *Geochim. Cosmochim. Acta*, 26 (1962) 225.
36 M. Florkin, Ch. Grégoire, S. Bricteux-Grégoire and E. Schoffeniels, *Compt. Rend.*, 252 (1961) 440.
37 R. R. Thompson and W. B. Creath, *Geochim. Cosmochim. Acta*, 30 (1966) 1137.
39 E. Abderhalden and K. Heyns, *Biochem. Z.*, 359 (1933) 320.
39 S. K. Das, A. R. Doberenz and R. W. G. Wyckoff, *Comp. Biochem. Physiol.*, 23 (1967) 519.
40 J. M. Everts, A. R. Doberenz and R. W. G. Wyckoff, *Comp. Biochem. Physiol.*, 26 (1968) 955.
41 R. W. G. Wyckoff, *The Biochemistry of Animal Fossils*, Williams and Wilkins, Baltimore, 1972.
42 J. D. Jones and J. R. Vallentyne, *Geochim. Cosmochim. Acta*, 21 (1960) 1.
43 S. Bricteux-Grégoire, M. Florkin and C. Grégoire, *Comp. Biochem. Physiol.*, 24 (1968) 567.

44 W. A. Isaacs, K. Little, J. D. Currey and L. B. H. Tarlo, *Nature*, 197 (1963) 192.
45 K. Little, M. Kelly and A. Courts, *Bone and Joint Surgery*, 44B (1962) 503.
46 R. W. G. Wyckoff, E. Wagner, R. Matter III and A. R. Doberenz, *Proc. Natl. Acad. Sci. (U.S.)*, 50 (1963) 215.
47 J. M. Shackleford and R. W. G. Wyckoff, *J. Ultrastruct. Res.*, 11 (1964) 173.
48 R. W. G. Wyckoff and A. R. Doberenz, *Proc. Natl. Acad. Sci. (U.S.)*, 53 (1965) 230.
49 R. W. G. Wyckoff and A. R. Doberenz, *J. Microsc.*, 4 (1965) 271.
50 R. Pawlicki, A. Korbel and H. Kubiak, *Nature*, 211 (1966) 655.
51 A. R. Doberenz and R. Lund, *Nature*, 212 (1966) 1502.
52 A. R. Doberenz and R. W. G. Wyckoff, *Proc. Natl. Acad. Sci. (U.S.)*, 57 (1967) 539.
53 T. Y. Ho, *Proc. Natl. Acad. Sci. (U.S.)*, 54 (1965) 26.
54 T. Y. Ho, *Comp. Biochem. Physiol.*, 18 (1966) 353.
55 T. Y. Ho, *Biochim. Biophys. Acta*, 133 (1967) 568.
56 M. F. Miller and R. W. G. Wyckoff, *Proc. Natl. Acad. Sci. (U.S.)*, 60 (1968) 176.
57 M. F. Voss-Foucart, *Comp. Biochem. Physiol.*, 24 (1968) 31.
58 M. F. Foucart, S. Bricteux-Grégoire, C. Jeuniaux and M. Florkin, *Life Sci.*, 4 (1965) 467.
59 M. Jope, *Comp. Biochem. Physiol.*, 20 (1967) 601.
60 M. Jope, *Comp. Biochem. Physiol.*, 30 (1969) 225.
61 R. M. Mitterer, *Ph. D. Geology*, Florida State Univ., 1966.
62 J. Grandjean, C. Grégoire and A. Lutts, *Bull. Classe Sci., Acad. Roy. Belg.*, 50 (1964) 562.
63 A. Hall and W. J. Kennedy, *Proc. Roy. Soc. (London), Ser. B*, 168 (1967) 377.
64 J. D. Hudson, *Geochim. Cosmochim. Acta*, 31 (1967) 2361.
65 C. Grégoire, *Nature*, 203 (1964) 868.
66 J. Halver and W. Shanks, *J. Nutr.*, 72 (1960) 340.
67 R. A. Coulson and T. Hernandez, in: J. W. Campbell (Ed.), *Comparative Biochemistry of Nitrogen Metabolism*, Vol. 2, Academic Press, New York, 1970.
68 G. W. Brown Jr., in: J. W. Campbell (Ed.), *Comparative Biochemistry of Nitrogen Metabolism*, Vol. 2, Academic Press, New York, 1970.
69 C. W. Scull and W. C. Rose, *J. Biol. Chem.*, 89 (1930) 109.
70 R. B. Band, *Nature*, 192 (1961) 674.
71 K. M. G. Adam, *J. Protozool.*, 11 (1964) 98.
72 G. W. Kidder, in: *Chemical Zoology*, Vol. 1 (see ref. 1).
72a W. M. Fitch, *Systematic Zool.*, 19 (1970) 99.
73 M. Florkin, in: *Chemical Zoology*, Vol. 1 (see ref. 1).
74 Ch. Jeuniaux, *Chitine et Chitinolyse*, Masson, Paris, 1963.
75 G. Ubaghs, in: *Chemical Zoology*, Vol. 3 (see ref. 1).
76 W. F. Neuman and M. W. Neuman, *Chem. Rev.*, 53 (1953) 1.
77 M. G. Dobb, R. D. B. Fraser and T. P. Macrae, *J. Cell Biol.*, 32 (1967) 297.
78 M. Ottensen, *Ann. Rev. Biochem.*, 36 (1967) 55.
79 R. A. Dickerson and I. Geis, *The Structure and Action of Proteins*, Harper and Row, New York, 1969.
80 V. Tomasek, F. Sorm, R. Zwilling and G. Preidlerer, *FEBS Letters*, 6 (1970) 229.
81 D. Gibson and G. H. Dixon, *Nature*, 222 (1969) 753.
82 T. H. Jukes, *Molecules and Evolution*, Columbia Univ. Press, New York, 1966.
83 H. Neurath and K. A. Walsh, *FEBS Letters*, 14 (1971) 222.
84 H. Neurath and G. H. Dixon, *Federation Proc.*, 16 (1957) 791.
85 M. Delaage, P. Desnuelle, M. Ladzuniski, E. Bricas and J. Savrda, *Biochem. Biophys. Res. Commun.*, 29 (1967) 235.
86 J. P. Abita, M. Delaage, M. Ladzunski and J. Savrda, *Europ. J. Biochem.*, 8 (1969) 314.
87 R. A. Bradshaw, H. Neurath, R. W. Tye, K. A. Walsh and W. P. Winter, *Nature*, 22 (1970) 237.

88   C. I. Harris and T. Hofmann, *Biochem. J.*, 114 (1969) 82P.
89   S. Bricteux-Grégoire, R. Schyns and M. Florkin, *Arch. Intern. Physiol. Biochem.*, 77 (1969) 544.
90   S. Bricteux-Grégoire, R. Schyns and M. Florkin, *Biochim. Biophys. Acta*, 351 (1974) 87.
91   M. Charles, M. Rovery, A. Guidoni and P. Desnuelle, *Biochim. Biophys. Acta*, 69 (1963) 115.
92   S. Bricteux-Grégoire, R. Schyns and M. Florkin, *Biochim. Biophys. Acta*, 251 (1971) 79.
93   S. Bricteux-Grégoire, R. Schyns and M. Florkin, in: E. Schoffeniels (Ed.), *Biochemical Evolution and the Origin of Life*, North Holland, Amsterdam, 1971.
94   R. Schyns, S. Bricteux-Grégoire and M. Florkin, *Biochim. Biophys. Acta*, 175 (1969) 97.
95   S. Bricteux-Grégoire, R. Schyns and M. Florkin, *Biochim. Biophys. Acta*, 229 (1971) 123.
96   E. W. Davie and H. Neurath, *J. Biol. Chem.*, 212 (1955) 515.
97   A. Puigserver and P. Desnuelle, *Biochim. Biophys. Acta*, 236 (1971) 499.
98   P. Desnuelle, J. Baratty and M. Rovery, in: P. Desnuelle, H. Neurath and M. Ottesen (Eds.), *Structure–Function Relationships of Proteolytic Enzymes*, Munksgaard, Copenhagen, 1970.
99   P. Voytek and E. C. Gjessing, *J. Biol. Chem.*, 246 (1971) 508.
100   G. R. Reeck and H. Neurath, *Biochemistry*, 11 (1972) 503.
101   G. Duchâteau and M. Florkin, *Bull. Soc. Chim. Biol.*, (1954) 295.
102   T. J. Bowen and B. A. Kilby, *Arch. Intern. Physiol. Biochim.*, 61 (1953) 413.
103   W. B. Hardy, *J. Physiol. (London)*, 13 (1892) 165.
104   W. D. Halliburton, *J. Physiol. (London)*, 6 (1885) 300.
105   J. Tait, *J. Physiol. (London)*, 40 (1910) XLI.
106   J. Tait and J. D. Gunn, *Quart. J. Exptl. Physiol.*, 12 (1918) 35.
107   Ch. Grégoire and M. Florkin, *Physiol. Comp. Oecol.*, 2 (1950) 126.
108   J. F. Yeager and H. H. Knight, *Ann. Entomol. Soc. Am.*, 26 (1933) 591.
109   Ch. Grégoire, in: *Chemical Zoology*, Vol. 6 (see ref. 1).
110   R. F. Doolittle and B. Blombäck, *Nature*, 202 (1964) 147.
111   B. Blombäck, in: E. Schoffeniels (Ed.) *Biochemical Evolution and the Origin of Life*, North-Holland, Amsterdam, 1971.
112   R. E. Dickerson, *J. Mol. Evol.*, 1 (1971) 26.
113   M. M. Guest, *Federation Proc.*, 25 (I) (1966) 73.
114   G. Kartha, J. Bello and D. Harker, *Nature*, 213 (1967) 862.
115   C. A. Long, *Bioscience*, 19 (1969) 519.
116   S. Basu, B. Kaufman and S. Roseman, *J. Biol. Chem.*, 240 (1965) PC 4115.
117   A. Gottschalk, *Nature*, 222 (1969) 452.
118   E. J. McGuire, G. M. Jourdian, D. M. Calson and S. Roseman, *J. Biol. Chem.*, 240 (1965) PC 4112.
119   W. T. Morgan and W. M. Watkins, *Brit. Med. Bull.*, 25 (1969) 30.
120   U. Brodbeck and K. E. Ebner, *J. Biol. Chem.*, 241 (1966) 762.
121   U. Brodbeck, W. L. Denton, N. Tanahashi and K. E. Ebner, *J. Biol. Chem.*, 242 (1967) 1391.
122   K. Brew, T. C. Vanaman and R. L. Hill, *Proc. Natl. Acad. Sci. (U.S.)*, 59 (1968) 491.
123   K. Brew, *Essays Biochem.*, 6 (1970) 93.
124   D. B. Carlisle, *Nature*, 172 (1953) 1098.
125   J. G. Pierce and T. H. Liao, *Federation Proc.*, 29 (1970) 600.
126   T. H. Liao and J. G. Pierce, *J. Biol. Chem.*, 245 (1970) 3275.
127   W. R. Butt and J. F. Kennedy, in: M. Margoulies and F. C. Greenwood (Eds.), *Protein and Polypeptide Hormones*, Part 1, Excerpta Medica, Amsterdam, 1971.
128   V. du Vigneaud, *Harvey Lectures*, Ser. 50 (1954–1955) 1.
129   R. Acher and C. Fromageot, *Ergeb. Physiol. Biol. Chem. Exptl. Pharmakol.*, 48 (1955) 286.
130   R. Archer, in: Comparative Aspects of Neurohypophyseal Morphology and Function, *Symp. Zool. Soc. (London)*, 9 (1963) 83.

131 R. Acher, J. Chauvet, M. T. Chauvet and D. Crepy, *Comp. Biochem. Physiol.*, 14 (1965) 245.
132 R. Acher, J. Chauvet, M. T. Chauvet and D. Crepy, *Biochim. Biophys. Acta*, 107 (1965) 393.
133 R. Acher, J. Chauvet and M. T. Chauvet, *Compt. Rend.*, 274 (1972) 313.
134 R. Acher, *Biochimie*, 56 (1974) 1.
135 R. L. Hill, *Federation Proc.*, 23 (1964) 1236.
136 M. F. Perutz, J. C. Kendrew and H. C. Watson, *J. Mol. Biol.*, 13 (1965) 669.
137 H. A. Itano, *Advan. Protein Chem.*, 12 (1957) 215.
138 V. M. Ingram, *Nature*, 189 (1961) 704.
139 J. Salkind, *Arch. Zool. Exptl. Gen.*, 55 (1915) 81.
140 H. M. Grey, *Advan. Immunol.*, 10 (1969) 51.
141 L. E. Hood, W. R. Gray, B. G. Sanders and W. J. Dreyer, *Cold Spring Harbor Symp. Quant. Biol.*, 32 (1967) 133.
142 S. Cohen and C. Milstein, *Advan. Immunol.*, 7 (1967) 1.
143 L. W. Clemm and P. A. Small, *J. Exptl. Med.*, 125 (1967) 893.
144 J. Marchalonis and G. M. Edelman, *J. Exptl. Med.*, 124 (1966) 901.
145 K. Titani, E. Whitly and F. W. Putman, *Science*, 152 (1966) 1513.
146 W. R. Gray, *Proc. Roy. Soc. (London)*, Ser. B, 166 (1966) 146.
147 S. J. Singer and R. F. Doolittle, *Science*, 153 (1966) 13.
148 H. M. Grey, in: E. Schoffeniels (Ed.), *Biochemical Evolution and the Origin of Life*, North-Holland, Amsterdam, 1971.
149 A. A. Suran and B. Papermaster, *Proc. Natl. Acad. Sci. (U.S.)*, 58 (1967) 1619.
150 G. M. Penn, H. G. Kunkel and H. M. Grey, *Federation Proc.*, 29 (2) (1970) 258
151 H. Tamir and S. Ratner, *Arch. Biochem. Biophys.*, 102 (1963) 249, 259.
152 G. W. Brown Jr. and P. P. Cohen, *Biochem. J.*, 75 (1960) 82.
153 S. R. R. Eddy and J. W. Campbell, *Comp. Biochem. Physiol.*, 28 (1969) 515.
154 S. R. R. Eddy and J. W. Campbell, *Biochem. J.*, 115 (1969) 495.
155 E. J. Conway and R. Cooke, *Biochem. J.*, 33 (1939) 457.
156 M. Florkin, *Arch. Intern. Physiol.*, 53 (1943) 117.
157 M. Florkin and H. Renwart, *Arch. Intern. Physiol.*, 49 (1939) 127.
158 M. Florkin and G. Frappez, *Arch. Intern. Physiol.*, 50 (1940) 197.
159 H. Delaunay, *Biol. Rev. Cambridge Phil. Soc.*, 6 (1931) 265.
160 W. T. W. Potts, *Comp. Biochem. Physiol.*, 14 (1965) 339.
161 I. Ehrlich, M. Rijavec and B. Kurelec, *Bull. Sci. Conseil Acad. RPF Yougoslavie*, 8 (1963) 133.
162 B. Kurelec, *Vet. Arhiv*, 34 (1964) 193.
163 M. M. Goil, *J. Helminthol.*, 32 (1958) 119.
164 J. W. Campbell and T. W. Lee, *Comp. Biochem. Physiol.*, 8 (1963) 29.
165 R. Cavier and J. Savel, *Bull. Soc. Chim. Biol.*, 36 (1954) 1631.
166 W. P. Rogers, *Australian J. Sci. Res.*, B5 (1952) 210.
167 S. Cohen and H. B. Lewis, *J. Biol. Chem.*, 180 (1949) 79.
168 A. E. Needham, *J. Exptl. Biol.*, 34 (1957) 425.
169 S. H. Bishop and J. W. Campbell, *Comp. Biochem. Physiol.*, 15 (1965) 51.
170 J. W. Campbell and S. H. Bishop, in: J. W. Campbell (Ed.), *Comparative Biochemistry of Nitrogen Metabolism*, Vol. 1, Academic Press, New York, 1970.
171 K. V. Speeg Jr. and J. W. Campbell, *Comp. Biochem. Physiol.*, 26 (1968) 579.
172 G. W. Brown Jr. and P. P. Cohen, *Biochem. J.*, 75 (1960) 89.
173 A. K. Huggins, G. Skutsch and E. Baldwin, *Comp. Biochem. Physiol.*, 28 (1969) 587.
174 L. J. Read, *Nature*, 215 (1967) 1412.
175 P. A. Janssens and P. P. Cohen, *Science*, 152 (1966) 358.

176   L. Goldstein, S. Harley-De Witt and R. P. Forster, *Comp. Biochem. Physiol.*, 44 B (1973)
      357.
177   E. Baldwin, *Nature*, 181 (1958) 1591.
178   A. D. Anderson and M. E. Jones, *Abstr(135th Meet. Amer. Chem. Soc., Div. Biol. Chem.*,
      No. 126 (1959).
179   L. J. Read, *Comp. Biochem. Physiol.*, 26 (1968) 455.
180   J. B. Balinsky, in: J. W. Campbell (Ed.), *Comparative Biochemistry of Nitrogen Metabolism*,
      Vol. 2, Academic Press, New York, 1970.
181   J. Needham, J. Brachet and R. K. Brown, *J. Exptl. Biol.*, 12 (1935) 321.
182   R. L. Watts and D. C. Watts, in: *Chemical Zoology*, Vol. 8 (see ref. 1).
183   H. W. Smith, *Biol. Rev. Cambridge Phil. Soc.*, 11 (1936) 49.
184   W. T. Neill, *Bull. Marine Sci. Gulf Caribbean*, 8 (1958) 1.
185   M. S. Gordon, K. Schmidt-Nielsen and H. M. Kelly, *J. Exptl. Biol.*, 38 (1961) 659.
186   K. Schmidt-Nielsen and P. Lee, *J. Exptl. Biol.*, 39 (1962) 167.
187   S. Thesleff and K. Schmidt-Nielsen, *J. Cellular Comp. Physiol.*, 59 (1962) 31.
188   E. Schoffeniels and R. R. Tercafs, *Ann. Soc. Roy. Zool. Belg.*, 96 (1966) 23.
189   M. S. Gordon, *J. Exptl. Biol.*, 39 (1962) 261.
190   B. T. Scheer and R. P. Markel, *Comp. Biochem. Physiol.*, 7 (1962) 289.
191   J. B. Balinsky, M. M. Cragg and E. Baldwin, *Comp. Biochem. Physiol.*, 3 (1961) 236.
192   M. Gilles-Baillien and E. Schoffeniels, *Ann. Soc. Roy. Zool. Belg.*, 95 (1965) 75.
193   Y. Charnot, *Trav. Inst. Sci. Chériffien, Sér. Zool.*, 20 (1960) 1.
194   M. Florkin, *Introduction Biochimique à la Chirurgie*, Masson, Paris, 1962.
195   A. E. Needham, in: J. W. Campbell (Ed.), *Comparative Biochemistry of Nitrogen Meta-
      bolism*, Vol. 1, Academic Press, New York, 1970.
196   R. R. Mills, C. R. Lake and W. L. Alworth, *J. Insect Physiol.*, 13 (1967) 1539.
197   E. J. W. Barrington and A. Thorpe, *Proc. Roy. Soc. (London)*, Ser. B, 163 (1965) 136.
198   I. Covelli, G. Salvatore, L. Sena and J. Roche, *Compt. Rend. Soc. Biol.*, 154 (1960) 1165.
199   J. Roche, G. Salvatore, G. Rametta and S. Varrone, *Compt. Rend. Soc. Biol.*, 153 (1959)
      1751.
200   W. Tong, P. Kerkof and I. L. Chaikoff, *Biochim. Biophys. Acta*, 56 (1962) 326.
201   A. Gorbman, M. Clements and R. O'Brien, *J. Exptl. Zool.*, 127 (1954) 75.
202   J. Roche, *Publ. Staz. Zool. Napoli*, Suppl. 31 (1959) 176.
203   J. Roche, S. André and I. Covelli, *Compt. Rend. Soc. Biol.*, 154 (1960) 2201.
204   W. Tong and I. L. Chaikoff, *Biochim. Biophys. Acta*, 48 (1961) 347.
205   C. W. Mayor, J. L. Hanegan and L. Anoli, *Comp. Biochem. Physiol.*, 28 (1969) 1153.
206   E. J. W. Barrington, in: E. Schoffeniels (Ed.), *Biochemical Evolution and the Origin of
      Life*, North-Holland, Amsterdam, 1971.
207   Z. M. Bacq and M. Florkin, in: M. J. Michelson (Ed.), *Comparative Pharmacology*, Vol. I,
      Pergamon, Oxford, 1973.
208   U. S. von Euler, in: U. S. von Euler and H. Heller (Eds.), *Comparative Endocrinology*,
      Vol. 2, Academic Press, New York, 1963.
209   J. F. Gaskell, *Phil. Trans. Roy. Soc. (London)*, Ser. B, 205 (1914) 153.
210   E. Ostlund, *Acta Physiol. Scand.*, 31, Suppl. 112 (1954) 1.
211   A. Bertler and E. Rosengren, *Acta Physiol. Scand.*, 47 (1959) 350.
212   G. A. Kerkut and R. J. Walker, *Comp. Biochem. Physiol.*, 3 (1961) 143.
213   M. J. Greenberg, *Brit. J. Pharmacol.*, 15 (1960) 365.
214   D. Sweeney, *Science*, 139 (1963) 1051.
215   G. Y. Kennedy, in: *Chemical Zoology*, Vol. 4 (see ref. 1).
216   L. Fredericq, *Bull. Acad. Roy. Belg.*, 35 (1898) 831.
217   E. Rodier, *Compt. Rend.*, 131 (1900) 1008.
218   M. N. Camien, H. Sarlet, Gh. Duchâteau and M. Florkin, *J. Biol. Chem.*, 193 (1951) 881.

219 Ch. Jeuniaux, Gh. Duchâteau-Bosson and M. Florkin, *J. Biochem. (Tokyo)*, 49 (1961) 527.
220 Gh. Duchâteau-Bosson, Ch. Jeuniaux and M. Florkin, *Arch. Intern. Physiol. Biochim.*, 69 (1961) 30.
221 Ch. Jeuniaux, S. Bricteux-Grégoire and M. Florkin, *Cahiers Biol. Mar.*, 3 (1962) 107.
222 Ch. Jeuniaux, S. Bricteux-Grégoire and M. Florkin, *Cahiers Biol. Mar.*, 2 (1961) 373.
223 Gh. Duchâteau, M. Florkin and Ch. Jeuniaux, *Arch. Intern. Physiol. Biochim.*, 67 (1959) 489.
224 S. Bricteux-Grégoire, Gh. Duchâteau-Bosson, Ch. Jeuniaux and M. Florkin, *Arch. Intern. Physiol. Biochim.*, 70 (1962) 273.
225 Gh. Duchâteau-Bosson and M. Florkin, *Comp. Biochem. Physiol.*, 3 (1961) 245.
226 Gh. Duchâteau, H. Sarlet, M. N. Camien and M. Florkin, *Arch. Intern. Physiol.*, 60 (1952) 124.
227 Gh. Duchâteau and M. Florkin, *Arch. Intern. Physiol.*, 62 (1954) 205.
228 H. H. Tallan, S. Moore and W. H. Stein, *J. Biol. Chem.*, 211 (1954) 927.
229 E. F. Beach, B. Munks and A. Robinson, *J. Biol. Chem.*, 148 (1943) 431.
230 R. J. Block, *Ann. N.Y. Acad. Sci.*, 53 (1951) 608.
231 A. P. M. Lockwood and P. C. Croghan, *Nature*, 184 (1959) 370.
232 D. W. Sutcliffe, *J. Exptl. Biol.*, 39 (1962) 325.
233 D. W. Sutcliffe, *Comp. Biochem. Physiol.*, 9 (1963) 121.
234 Ch. Jeuniaux, in: *Chemical Zoology*, Vol. 6 (see ref. 1).
235 S. Bricteux-Grégoire and M. Florkin, *Arch. Intern. Physiol. Biochim.*, 67 (1959) 29.
236 L. Levenbook, *J. Insect Physiol.*, 8 (1962) 559.
237 Gh. Duchâteau and M. Florkin, *Arch. Intern. Physiol. Biochim.*, 66 (1958) 573.
238 M. Florkin and Ch. Jeuniaux, in: M. Rockstein (Ed.), *The Physiology of Insecta*, 2nd edn., Vol. 5, Academic Press, New York, 1974.
239 R. Pant and H. C. Agrawal, *J. Insect Physiol.*, 10 (1964) 443.
240 A. Mansingh, *J. Insect Physiol.*, 13 (1967) 1645.
241 H. Brunnert, *Zool. Jahrb., Abt. Allgem. Zool. Tiere*, 73 (1967) 702.
242 C. M. Williams, *Harvey Lectures*, Ser. 47 (1951–1952) 126.
243 S. Bricteux-Grégoire, Gh. Duchâteau-Bosson, M. Florkin and W. G. Verly, *Arch. Intern. Physiol. Biochim.*, 68 (1960) 424.
244 J. Shaw, in: M. Florkin and H. S. Mason (Eds.), *Comparative Biochemistry*, Vol. 2, Academic Press, New York, 1960.
245 Gh. Duchâteau and M. Florkin, *Arch. Intern. Physiol. Biochim.*, 63 (1955) 249.
246 E. L. Gasteiger, P. C. Hoake and J. A. Gergen, *Ann N.Y. Acad. Sci.*, 90 (1960) 622.
247 J. R. Beers, *Comp. Biochem. Physiol.*, 21 (1967) 11.
248 W. Scholles, *Z. Vergleich. Physiol.*, 19 (1933) 522.
249 M. Florkin, *Colloq. Ges. Physiol. Chem.*, 6 (1956) 62.
250 M. Florkin, *Bull. Acad. Roy. Belg., Cl. Sci.*, 48 (1962) 687.
251 Gh. Duchâteau and M. Florkin, *J. Physiol. (Paris)*, 48 (1956) 520.
252 J. Shaw, *J. Exptl. Biol.*, 35 (1958) 902.
253 J. Shaw, *J. Exptl. Biol.*, 35 (1958) 920.
254 S. Bricteux-Grégoire, Gh. Duchâteau-Bosson, Ch. Jeuniaux and M. Florkin, *Comp. Biochem. Physiol.*, 19 (1966) 729.
255 W. T. W. Potts, *J. Exptl. Biol.*, 35 (1958) 749.
256 R. Lange, *Comp. Biochem. Physiol.*, 10 (1963) 173.
257 S. Bricteux-Grégoire, Gh. Duchâteau-Bosson, Ch. Jeuniaux and M. Florkin, *Arch. Intern. Physiol. Biochim.*, 72 (1964) 116.
258 S. Bricteux-Grégoire, Gh. Duchâteau-Bosson, Ch. Jeuniaux and M. Florkin, *Arch. Intern. Physiol. Biochim.*, 72 (1964) 835.

259 S. Bricteux-Grégoire, Gh. Duchâteau-Bosson, Ch. Jeuniaux and M. Florkin, *Arch. Intern. Physiol. Biochim.*, 72 (1964) 267.

260 K. Allen, *Biol. Bull.*, 121 (1961) 419.

261 M. P. Lynch and L. Wood, *Comp. Biochem. Physiol.*, 19 (1966) 783.

262 M. B. Peterson and F. G. Duerr, *Comp. Biochem. Physiol.*, 28 (1969) 633.

263 E. Schoffeniels and R. Gilles, in: *Chemical Zoology*, Vol. 5, Chapter 9 (see ref. 1).

264 R. A. Virkar and K. L. Webb, *Am. Zoologist*, 7 (1967) 735.

265 R. Lange, *Comp. Biochem. Physiol.*, 13 (1964) 205.

266 G. C. Stephens and R. A. Virkar, *Biol. Bull.*, 131 (1966) 172.

267 R. A. Virkar, *Comp. Biochem. Physiol.*, 18 (1966) 617.

268 J. Binyon, *J. Marine Biol. Assoc., U.K.*, 41 (1961) 161.

269 C. Schlieper, *Verhandl. Deut. Zool. Ges.*, 33 (1929) 214.

270 L. C. Beadle, *J. Exptl. Biol.*, 14 (1937) 56.

271 H. Brattsbröm, *Undersöknigar over Öresund*, Lund, 1941, p. 27.

272 S. Segersträle, *Oikos*, 1 (1949) 127.

273 C. Schlieper, *Année Biol.*, 33 (1957) 117.

274 J. Hoyaux, *Euryhalinité et Osmorégulation chez Quelques Gastéropodes de la Zone Intertidale*, Mém. de Licence en Sci. Zool., Liège, 1967.

275 R. Gilles and E. Schoffeniels, *Biochim. Biophys. Acta*, 82 (1964) 518.

276 M. Florkin, Gh. Duchâteau-Bosson, Ch. Jeuniaux and E. Schoffeniels, *Arch. Intern. Physiol. Biochim.*, 72 (1964) 892.

277 A. E. Needham, *Physiol. Comp. Oecol.*, 4 (1957) 209.

278 Ch. Jeuniaux and M. Florkin, *Arch. Intern. Physiol. Biochim.*, 69 (1961) 385.

279 E. Schoffeniels, *Arch. Intern. Physiol. Biochim.*, 68 (1960) 696.

280 R. Gilles and E. Schoffeniels, *Comp. Biochem. Physiol.*, 31 (1969) 927.

281 E. Schoffeniels, *Cellular Aspects of Membrane Permeability*, Pergamon, Oxford, 1957.

282 R. Gilles and E. Schoffeniels, *Bull. Soc. Chim. Biol.*, 48 (1966) 397.

283 R. Gilles, *Arch. Intern. Physiol. Biochim.*, 77 (1969) 441.

284 E. Schoffeniels, *Arch. Intern. Physiol. Biochim.*, 76 (1968) 319.

285 D. J. Candy and B. A. Kilby, *J. Exptl. Biol.*, 39 (1962) 129.

286 E. Jaworsky, L. Wang and G. Marco, *Nature*, 198 (1963) 790.

287 Ch. Jeuniaux, in: E. Schoffeniels (Ed.), *Biochemical Evolution and the Origin of Life*, North-Holland, Amsterdam, 1971.

288 G. Dandrifosse, *Ann. Soc. Roy. Zool. Belg.*, 92 (1962) 199.

289 C. F. Howden and B. A. Kilby, *J. Insect Physiol.*, 6 (1961) 85.

290 G. R. Wyatt and G. F. Kalf, *Federation Proc.*, 15 (1956) 388.

291 G. R. Wyatt and G. F. Kalf, *J. Gen. Physiol.*, 40 (1957) 833.

292 D. R. Evans and V. G. Dethier, *J. Insect Physiol.*, 1 (1957) 3.

293 L. J. Wimer, *Comp. Biochem. Physiol.*, 29 (1969) 1055.

294 M. Crompton and L. M. Birt, *J. Insect Physiol.*, 13 (1967) 1575.

295 S. Friedman, in: *Chemical Zoology*, Vol. 5 (see ref. 1).

296 I. Mochnacka and C. Petryszyn, *Acta Biochim. Polon.*, 6 (1959) 307.

297 D. Fairbairn, *Can. J. Zool.*, 36 (1858) 787.

298 C. von Czarnowsky, *Naturwissenschaften*, 41 (1954) 577.

299 J. S. Barlow and H. L. House, *J. Insect Physiol.*, 5 (1960) 1951.

300 L. Levenbook, *Nature*, 160 (1947) 465.

301 L. Levenbook, *Biochem. J.*, 47 (1950) 336.

302 A. Maurizio, *J. Insect Physiol.*, 11 (1965) 745.

303 F. G. Chang and S. Friedmann, in ref. 296.

304 D. J. Candy and B. A. Kilby, *Biochem. J.*, 78 (1961) 531.

305 G. Mikolashek and E. Zebe, *J. Insect Physiol.*, 13 (1967) 1483,

222

306 Gh. Duchâteau-Bosson, Ch. Jeuniaux and M. Florkin, *Arch. Intern. Physiol. Biochim.*, 73 (1963) 566.
307 S. Saito, *J. Insect Physiol.*, 9 (1963) 509.
308 J. E. Steele, *Nature*, 192 (1961) 680.
309 T. A. Murphy and G. R. Wyatt, *J. Biol. Chem.*, 240 (1965) 1500.
310 C. l'Hélias, *Compt. Rend.*, 236 (1953) 2489.
311 R. McCarthy and C. L. Ralph, *Am. Zoologist*, 2 (1962) 429.
312 J. E. Steele, *J. Insect Physiol.*, 15 (1969) 421.
313 F. Stanley, *J. Insect Physiol.*, 12 (1966) 397.
314 A. Gelperin, *J. Insect Physiol.*, 12 (1966) 331.
315 E. Thomsen, *J. Exptl. Biol.*, 29 (1952) 137.
316 M. Kobayashi and S. Kirimura, *J. Insect Physiol.*, 13 (1967) 545.
317 J. E. Steele, *Gen. Comp. Endocrinol.*, 3 (1963) 46.
318 S. Bricteux-Grégoire, Ch. Jeuniaux and M. Florkin, *Arch. Intern. Physiol. Biochim.*, 72 (1964) 482.
319 S. Bricteux-Grégoire, Ch. Jeuniaux and M. Florkin, *Comp. Biochem. Physiol.*, 16 (1965) 333.
320 T. A. Murphy and G. R. Wyatt, *Nature*, 202 (1964) 1112.
321 G. F. Kalf and D. V. Rieder, *J. Biol. Chem.*, 230 (1958) 691.
322 G. F. Howden and B. A. Kilby, *Chem. Ind. (London)*, (1956) 1453.
323 E. C. Zebe and W. H. McShan, *J. Cellular Comp. Physiol.*, 53 (1959) 21.
324 S. Friedman, *Arch. Biochem. Biophys.*, 93 (1961) 550.
325 A. E. S. Gussin and G. R. Wyatt, *Arch. Biochem. Biophys.*, 112 (1965) 626.
326 A. R. Gilby, S. S. Wyatt and G. R. Wyatt, *Acta Biochim. Polon.*, 14 (1967) 83.
327 E. Stevenson, *J. Insect Physiol.*, 14 (1968) 179.
328 K. Hansen, *Biochem. Z.*, 344 (1966) 15.
329 J. S. Clegg and D. R. Evans, *J. Exptl. Biol.*, 38 (1961) 771.
330 Th. Bücher and M. Klinkenberg, *Angew. Chem.*, 70 (1958) 552.
331 Gh. Duchâteau and M. Florkin, *Arch. Intern. Physiol. Biochim.*, 67 (1959) 306.
332 D. J. Candy and B. A. Kilby, *J. Exptl. Biol.*, 39 (1962) 129.
333 C. K. Chang, F. Liu and H. Feng, *Acta Entomol. Sinica*, 13 (1964) 494.
334 R. G. Hansford and B. Sacktor, in: *Chemical Zoology*, Vol. 6B (see ref. 1).
335 H. A. Schneiderman and C. M. Williams, *Biol. Bull.*, 105 (1953) 320.
336 H. G. Andrewartha, *Biol. Rev. Cambridge Phil. Soc.*, 27 (1952) 50.
337 R. W. Salt, *Can. J. Zool.*, 37 (1959) 59.
338 G. R. Wyatt and W. L. Meyer, *J. Gen. Physiol.*, 42 (1959) 1005.
339 R. C. Wilhelm, H. A. Schneiderman and L. J. Daniel, *J. Insect Physiol.*, 7 (1961) 273.
340 R. C. Wilhelm, *Ph. D. Thesis*, Cornell University, Ithaca, N.Y.
341 Y. Lin, J. G. Duman and A. L. de Vries, *Biochem. Biophys. Res. Commun.*, 46 (1972) 87.
342 J. S. Clegg, *J. Exptl. Biol.*, 41 (1964) 879.
343 B. Heinrich, *Science*, 175 (1972) 185.
344 E. A. Newsholme, B. Crabtree, S. J. Higgins, S. D. Thornton and C. Start, *Biochem. J.*, 128 (1972) 89.
345 G. J. Brewer and J. W. Eaton, *Science*, 171 (1971) 1205.
346 H. G. Hers, *Biochim. Biophys. Acta*, 37 (1960) 120, 127.
347 D. C. Jackson, *J. Appl. Physiol.*, 24 (1968) 503.
348 R. E. Smith, *Federation Proc.*, 19, Suppl. 5 (1960) 64.
349 F. Ismail-Beigi and I. S. Edelman, *Proc. Natl. Acad. Sci. (U.S.)*, 67 (1970) 1071.
350 G. Steiner, G. E. Johnson, E. A. Sellers and E. Schönbaum, *Federation Proc.*, 28 (1969) 1017.
351 P. T. Grant and J. R. Sargent, *Biochem. J.*, 76 (1960) 229.

352   R. W. Bide and P. T. Grant, *Abstr. 1st Meeting Feder. Europ. Biochem. Soc.*, London, 1964, p. 72.
353   K. Bloch, in: V. Bryson and H. J. Vogel (Eds.), *Evolving Genes and Proteins*, Academic Press, New York, 1965.
354   K. Bloch, in: C. A. Leone (Ed.), *Taxonomic Biochemistry and Serology*, Ronald, New York, 1964.
355   R. B. Clayton, *J. Lipid Res.*, 5 (1964) 3.
356   C. H. Schaefer, J. N. Kaplanis and W. E. Robbins, *J. Insect Physiol.*, 11 (1965) 1013.
357   J. D. Chanley, R. Mezzetti and H. Sobotka, *Tetrahedron*, 22 (1966) 1857.
358   G. Habermehl, *Nature*, 53 (1966) 123.
359   G. Habermehl and A. Hauf, *Chem. Ber.*, 101 (1968) 198.
360   E. Wölfel, C. Schöpf, G. Weitz and G. Habermehl, *Chem. Ber.*, 44 (1961) 2361.
361   G. Habermehl, *Chem. Ber.*, 96 (1963) 143.
362   G. Habermehl, *Chem. Ber.*, 99 (1966) 1439.
363   G. Habermehl, *Z. Naturforsch.*, 20 (1965) 1129.
364   M. D. Siperstein, A. W. Murray and E. Titius, *Arch. Biochem. Biophys.*, 67 (1957) 157.
365   H. Schildknecht and D. Hotz, *Angew. Chem., Intern. Edn. Engl.*, 6 (1967) 881.
366   H. Schildknecht, R. Siewerdt and U. Maschwitz, *Angew. Chem., Intern. Edn. Engl.*, 5 (1966) 421.
367   H. Schildknecht, H. Birringer and U. Maschwitz, *Angew. Chem., Intern. Edn. Engl.*, 6 (1967) 558.
368   R. B. Clayton, in: E. Sondheimer and J. B. Simeone (Eds.), *Chemical Ecology*, Academic Press, New York, 1970.
369   H. A. Bern, *Am. Zoologist*, 7 (1967) 815.
370   P. Karlson and H. Hoffmeister, *Z. Physiol. Chem.*, 331 (1963) 298.
371   P. Karlson, *Pure Appl. Chem.*, 14 (1967) 75.
372   D. S. King and J. B. Siddall, *Nature*, 221 (1969) 955.
373   J. A. Thomson, J. B. Siddall, M. N. Galbraith, D. H. S. Horn and E. J. Middleton, *Chem. Comm.*, (1969) 669.
374   J. B. Siddall, in: E. Sondheimer and J. B. Simeone (Eds.), *Chemical Ecology*, Academic Press, New York, 1970.
375   E. O. Wilson and W. H. Bossert, *Recent Progr. Hormone Res.*, 19 (1963) 673.
376   J. Weatherston, *Quart. Rev. (London)*, 21 (1967) 287.
377   F. E. Regnier and J. H. Law, *J. Lipid Res.*, 9 (1968) 541.
378   M. Florkin and E. Schoffeniels, *Molecular Approaches to Ecology*, Academic Press, New York, 1969.
379   G. M. Happ and J. Meinwald, *J. Am. Chem. Soc.*, 87 (1965) 2059.
380   J. Meinwald, G. M. Happ, J. Labows and T. Eisner, *Science*, 151 (1969) 79.
381   M. Pavan and R. Trave, *Insectes Sociaux*, 5 (1958) 299.
382   A. Quilico, P. Grünanger and M. Pavan, *Proc. 11th Intern. Congr. Entomol.*, Vienna, 1960, Vol. 3, 1962, p. 66.
383   G. W. K. Cavill and H. Hinterberger, *Proc. 11th Intern. Congr. Entomol.*, Vienna, 1960, Vol. 3, 1962, p. 53.
384   D. L. Wood, L. E. Browne, W. D. Bedard, P. E. Tilden, R. M. Silverstein and J. O. Rodin, *Science*, 159 (1968) 1373.
385   R. M. Silverstein, J. O. Rodin and D. L. Wood, *Science*, 154 (1966) 509.
386   G. Bergström, B. Kullenberg, S. Stälberg-Stenhagen and E. Stenhagen, *Arkiv Kemi*, 28 (1967) 453.
387   L. R. Cleveland, *Science*, 105 (1947) 16.
388   G. Wald, *Harvey Lectures*, Ser. 41 (1946) 117.
389   C. D. B. Bridges, *Cold Spring Harbor Symp. Quant. Biol.*, 30 (1965) 317.

390   H. J. A. Dartnall, in: H. Davson (Ed.), *The Eye*, Vol. 2, Academic Press, New York, 1922.
391   F. Crescitelli, in: *Photobiology, Proc. 19th Ann. Biol. Colloq., Corvallis, Oregon*, 1958.
392   R. A. Morton and G. A. J. Pitt, in: *Math. Phys. Lab. Symp.*, No. 8, H.M.S.O., London, 1958.
393   T. H. Goldsmith, *Ann. N.Y. Acad. Sci.*, 74 (1958) 266.
394   T. H. Goldsmith, *Proc. Natl. Acad. Sci. (U.S.)*, 44 (1958) 123.
395   E. J. Denton and J. H. Wyllie, *J. Physiol. (London)*, 127 (1955) 81.
396   K. O. Donner and T. Reuter, *Vision Res.*, 2 (1962) 357.
397   C. D. B. Bridges, in: M. Florkin and E. H. Stotz (Eds.), *Comprehensive Biochemistry*, Vol. 27, Elsevier, Amsterdam, 1967, p. 31.
398   G. Wald, P. K. Brown and P. S. Smith, *J. Gen. Physiol.*, 38 (1955) 623.
399   G. Wald, in: M. Florkin and H. S. Mason (Eds.), *Comparative Biochemistry*, Vol. 1, Academic Press, New York, 1960.
400   F. Crescitelli and H. J. A. Dartnall, *J. Physiol. (London)*, 125 (1954) 604.
401   C. D. B. Bridges, *Nature*, 203 (1964) 303.
402   F. Crescitelli, *Ann. N. Y. Acad. Sci.*, 74 (1958) 230.
403   C. D. B. Bridges, *Vision Res.*, 5 (1965) 239.
404   F. H. Wilt, *J. Embryol. Exptl. Morphol.*, 7 (1959) 556.
405   K. Naito and F. H. Wilt, *J. Biol. Chem.*, 237 (1962) 3060.
406   G. Wald, *J. Gen. Physiol.*, 25 (1941) 235.
407   E. M. Kampa, *J. Physiol. (London)*, 119 (1953) 400.
408   C. D. B. Bridges, *J. Physiol. (London)*, 134 (1956) 620.
409   F. W. Munz and D. D. Beatty, *Vision Res.*, 5 (1965) 1.
410   D. D. Beatty, *Can. J. Zool.*, 44 (1966) 429.
411   E. M. Kampa, B. C. Abbott and B. P. Boden, *J. Marine Biol. Assoc. U.K.*, 43 (1963) 683.
412   H. Stieve, *Z. Vergleich. Physiol.*, 43 (1960) 518.
413   F. W. Munz, in: G. E. Wolstenhome and J. Knight (Eds.), *Ciba Foundation Symposium on Physiology and Experimental Psychology of Colour Vision*, Churchill, London, 1965.
414   F. Crescitelli, *J. Gen. Physiol.*, 40 (1956) 217.

Chapter V

Ideas and Experiments in the Field of Prebiological Chemical Evolution

MARCEL FLORKIN

Department of Biochemistry, University of Liège (Belgium)

1. Heterogeny (spontaneous generation)

Spontaneous generation is an old belief which was already a part of the theories of the ancient Greek philosophers. Whether from mud and silt or from decaying parts of dead organisms, all organisms were believed to come into being as described by Shakespeare in *Anthony and Cleopatra:*

Your serpent of Egypt is bred now of your mud
By the operation of your sun: so is your crocodile

(Act III, Scene 7)

Van Helmont[1] offers a procedure for the spontaneous generation of mice by putting in an open jar a piece of underwear soiled with sweat, together with some wheat. After about three weeks, mice of both sexes emerge in the adult stage and they are able to reproduce with mice born from parents. The interpretation given by Van Helmont is that the "ferment" of the sweat, coming out of the piece of soiled underwear, has penetrated through the husks of the wheat and changed the wheat into mice. In the same text, Van Helmont recalls that body lice, bugs, fleas and worms are born "from our entrails and excrements".

Redi[2] in the course of the 17th century performed experiments which shook the theory of spontaneous generation. Having left a piece of snake's meat in an open box and noticing, after a few days, the presence of little worms on the meat, he isolated and observed these worms out of curiosity and saw them change into flies. He concluded that they were the larvae of insects.

References p. 259

[231]

The impact of this discovery must be appreciated in the context of the natural philosophy of the time, which was greatly concerned with Harvey's novel idea that animals are produced from eggs. Controversies were raging about the question whether animals come from organic parts or from eggs. To settle this question, Redi started experiments on maggots. He took two pieces of meat, wrapped one in muslin and put in it a container whose opening he also covered with muslin. The other piece was left in an open container. In the latter, after several days, the decaying meat was found covered with larvae and eggs while in the control meat wrapped in muslin, no eggs or larvae were found. From this Redi concluded that the laying of eggs was a necessary condition for the appearance of "worms" in the decaying meat. But the controversy started anew[3] when Van Leeuwenhoek, peering through his primitive microscope at a drop of Lake Berkelse water, observed it teeming with "animalcules". Réaumur and Bonnet opposed the theory that the animalcules arose by spontaneous generation while Buffon accepted it. The theory received apparent confirmation on the strength of observations by J. T. Needham. He put into glass flasks different infusions, boiled them for several minutes and corked the flasks. In all cases, after a few days, the infusions were teeming with "infusoria", a fact Needham and many followers took to be a demonstration of the spontaneous generation of "infusoria" from organic matter.

These experiments were criticized by Bonnet and by De Lignac and re-examined by Spallanzani[3]. He first sealed the vessels hermetically and boiled the contents for a long time, up to one hour. A few days after, when the infusions were subjected to microscopic examination, no animalcules were detected. This was confirmed later by Joblot[4].

These experiments failed to convince Needham[5] who retorted that the capacity of air to sustain life in the vials had been destroyed in Spallanzani's brutal treatment.

Needham[5] writes that Spallanzani

"hermetically sealed nineteen vessels filled with different vegetable substances and boiled them thus closed for the period of an hour. But from the method of treatment by which he had tortured his nineteen vegetable infusions, it is plain that he has greatly weakened, or perhaps entirely destroyed, the vegetative force of the infused substances. And, not only this, he has by the exhalations and by the intensity of fire, entirely spoiled the small amount of air that remained in the empty part of his vessels. Consequently it is not surprising that his infusions thus treated gave no sign of life. This is as it should have been."

Surprise is sometimes expressed at the ease with which the scientific community accepted this answer. But we must consider that the concept of spontaneous generation had nothing in common with our present concept of the self-association of molecular units of non-biological origin as the first form of precellular units. In the minds of the scientists of the mid-18th century, spontaneous generation was the result of the association of particles of *life-matter* and had no relation whatsoever with the "origin of life" from non-living substances, a concept which arose only in the 19th century. To conceive that boiling could jeopardize properties corresponds, *mutatis mutandis*, to our modern views on molecular information governing self-assembly; this was no form of stupidity but on the contrary a manifestation of deep insight.

Spallanzani[3] performed new experiments with the purpose of separating the effects of heat on the organic matter used in the experiments, and on the oxygen. He showed that when the experiment is made with an infusion of carbonized seeds, the infusoria still appear, a peremptory denial of the influence of the "vegetative force" of the seeds as stressed by Needham. At the time the composition of air was still unknown and Spallanzani's experiments on that issue failed to carry conviction. In a book published in 1805, one of the leaders of *Naturphilosophie*, Oken[6], again stated

"The genesis of infusoria is not due to a development from eggs but to the release of bonds within larger animals, a dislocation of the animal in its constitutive animals". (translation by author)

In 1809, Appert[7] in the course of inventing the canning process simply repeated Spallanzani's classical experiment. Studying the mechanism of Appert's process, Gay-Lussac[8] introduced the concept of oxygen as the effector of putrefaction.

The reader will find in Vol. 30 of this Treatise (Chapter 6) a review of Schwann's contribution to the analysis of the mechanism of putrefaction as due to microorganisms. Schwann started with tests to determine whether the properties of oxygen are indeed modified during the heating of the infusions, as Gay-Lussac held. His conclusion as stated in his doctoral thesis that "*Infusoria non oriunter generatione equivoca*" led him to suggest that putrefaction was the result of the activity of microorganisms and to discard the theory according to which the latter are generated from organic matter in the course of putrefaction. The result of Schwann's experiments was the refutation of Gay-Lussac's oxygen theory; his views on putrefaction and, indirectly, those on fermentation are derived from this conclusion.

The German physiologist Burdach[9] introduced the terms *homogeny* to designate the procreation of living creatures by parents, and *heterogeny* to designate non-parental procreation.

Heterogeny was synonymous with spontaneous generation, primordial generation or *generatio equivoca*. Edwards[10]*, in his lectures on physiology and comparative anatomy distinguishes several kinds of heterogeny. He speaks of an a-genetic heterogeny to designate generation from inorganic or organic matter without any participation of living creatures, for instance from CO_2, NH_3, H_2O etc., or from albumin, fibrin, cellulose, etc. Necrogenetic generation results from the dissociation of parts of organisms without loss of vital power. In xenogenetic generation, life is transmitted to inert matter by the physiological action of the "power" of a living organism.

Pasteur[12] attacked the doctrine of spontaneous generation in a classic paper which appeared in 1862. Pasteur had reached the conclusion that all "fermentations" were mediated by living organisms. He therefore followed Schwann in maintaining that either the air carried germs into the infusion (as Schwann had concluded) or, as adherents of the spontaneous generation theory believed, that the microorganisms of putrefaction were the result of the recombination of life-matter released from decaying organic matter. The choice was between putrefaction resulting from the development of germs, and the generation of germs in decaying organic matter after contact with oxygen.

If the former view is correct, the air must contain the germs as Schwann concluded. Pasteur resorted to Schwann's method of depriving the air of germs by the action of heat.

Pasteur's demonstration fell short of convincing his opponent Pouchet[13], who stuck to the view that what was modified in Pasteur's experiments, was the air itself. He boiled infusions of hay, sealed the flasks and observed no growth as long as the flasks were kept closed. If the seals were broken and a small amount of air allowed to penetrate, growth was observed upon re-sealing of the flasks, and this was the case whether the experiment was made in Paris or in the Pyrenees mountains.

These arguments were finally discarded by Tyndall[14] when he showed

---

* Henri-Milne Edwards, one of his father's 29 children, was the brother of W. F. Edwards, the physiologist and author of well-known studies on respiration. He is sometimes given the patronymic name Milne-Edwards, which is not acceptable practice as it was his son Alphonse who first adopted this name (see Théodoridès[11]).

that the germs of hay infusions survive boiling for long periods of time and grow in the presence of air.

It has been shown by Edwards[10] and later by Farley[15] that if Pasteur's experiments marked the end of the concepts of agenesis and necrogenesis, they did not abolish belief in xenogenetic generation. The theory of spontaneous generation did not die out until the discovery in the 1850's (by Küchenmeister, von Siebold, P. J. Van Beneden and Leuckart) of the life cycle of parasitic worms and of their migration through a series of hosts to complete their development stages (see Edwards[10] and Farley[15]).

Astonishingly enough, the charge of vitalism has occasionally been directed to Schwann and Pasteur because of their experiments designed to refute the theory of spontaneous generation. The vitalists were those who believed in necrogeny or xenogeny as the effects of a "vital force" or "vital power". If vitalism is the belief that all organisms now existing on earth are derived from other organisms, then all biologists living today are vitalists.

2. Agenesis* or the origin of life

Among the varieties of heterogenesis recorded we have mentioned (following Edwards' terminology[10]) the agenetic formation, or the generation from inorganic or organic matter without any participation of living beings or of their parts.

Agenesis (or more currently abiogenesis) as well as necrogenesis and xenogenesis were recognized as not involved in the generation of the organisms now living. But this does not exclude their participation in the "origin of life".

With respect to this distinction between "spontaneous generation" and "the origin of life", Pasteur has made clear his attitude by writing that

"If, as a result of (his) experimental work, (he) happened to demonstrate that matter can become organized of itself into a cell or living being", (cited by Fox[16])

he would be ready to recognize the self-organization of matter into a cell.

Pasteur's intellectual attitude is correctly described in the article on spontaneous generation contained in the *Nouveau Larousse Illustré*[17].

---

* This term was introduced by Edwards[10] and is used here in its later form (abiogenesis), in preference to other, less appropriate terms.

References p. 259

"Pasteur has shown that in the conditions of his experiments, it is an error to believe in the spontaneous generation of bacteria and moulds; never did he claim that, in other conditions, living beings simpler than those could not be formed directly at the expense of non-living material". (translation by author)

Pasteur, in a a letter to his opponent Pouchet, quoted by Mondor[18], wrote:

"I believe that you are wrong not when you believe in spontaneous generation... but when you affirm its occurrence. In the experimental sciences, it is always wrong not to doubt so long as the facts do not compel affirmation". (translation by author)

In his well-known paper on intracellular respiration and its molecular basis, Pflüger[18a] has formulated a theory of the origin of life in which he suggests the existence of an intermediary period (Zwischenstadium) leading "von der leblosen zur lebendigen Natur", *i.e.* what we call today chemical evolution.

To quote Pflüger,

"Der frische Stickstoff der Luft ist fähig, wenn er mit einem stark glühenden Gemenge von Kalium und Kohle oder mit einem bis zur Weissgluth erhitzten Gemenge von Kali oder kohlensaurem Kali und Kohle zusammenkommt, Cyan-Kalium zu bilden. — Die Sauerstoffverbindungen des Stickstoffs — Salpetersäure bildet sich z.B. ja bei Gewittern — liefern ferner unter ähnlichen Bedingungen weit leichter Cyanverbindungen. — Ferner: Ammoniak über glühende Kohle geleitet bildet Cyanammonium; ebenso ein Gemenge von Kohlenoxyd und Ammoniak in Berührung mit glühendem Platinschwamm. — Ferner: Wird Ammoniak über ein glühendes Gemenge von kohlensaurem Kali und Kohle geleitet, oder wird Salmiak mit kohlensaurem Kali und Kali geglüht, so erhält man Cyankalium. — Ferner: Wenn Kohlenoxydgas mit Kalihydrat längere Zeit erhitzt wird, bildet sich ameisensaures Kali, welches sich mit einem Ammoniumsalz in ameisensaures Ammonium umsetzen kann. Ameisensaures Ammonium liefert beim Erhitzen für sich oder mit wasserentziehenden Substanzen unter Verlust von Wasser: Cyanwasserstoff, Cyan oder Cyansäure.

"Es ist sonach nichts klarer, als die Möglichkeit der Bildung von Cyanverbindungen, als die Erde noch *ganz* oder *partiell* in feurigem oder erhitztem Zustande war. Ich stelle mir vor, man müsse daran denken, dass die Abkühlung auf der Erdoberfläche nicht gleichförmig geschah und dass einzelne Distrikte, die sich abgekühlt hatten, auch wieder erhitzt werden konnten u.s.w.

"Ebenso ist principiell zu begreifen, was kein Chemiker leugnen wird, die Entstehung der anderen wesentlichen Constituenten des Eiweissmolecüles, nämlich zahlloser Kohlenwasserstoffe, resp. Alkoholradicale ohne irgend welche Vermittelung lebendiger Materie durch synthetische Bildungen. Nachdem wir die Bedingungen der Synthese des Cyans kennen gelernt, fragen wir nach denen der Kohlenwasserstoffe.

"Wenn Schwefelkohlenstoffdampf mit Schwefelwasserstoff über glühende Metalle geleitet wird, so entsteht Aethylen. Wenn Schwefelkohlenstoff mit Schwefelwasserstoff oder auch mit Wasserdampf auf glühende Metalle geleitet wird, entsteht auch, wie Berthelot fand, Methylwasserstoff, Kohlenstoff und Wasserstoff vereinigen sich unter Mitwirkung electrischer

Entladungen zu Acetylen, und, dieses giebt mit Sauerstoff Oxalsäure. Beim Durchleiten einer Mischung von Methylwasserstoff und Kohlenoxydgas durch eine glühende Röhre erhält man Propylen. Bei Destillation von ameisensaurem Baryt wird nach Berthelot Sumpfgas, Aethylen und Propylen gebildet. Bei Destillation von ameisensaurem Natron, das sich leicht aus den Elementen erzeugt, mit Natronkalk entstehen Methylwasserstoff, Aethylen, Butylen, Amylen u.s.w.

"Da das Eiweiss sicher den Benzolkohlenkern enthält, weil durch einfache chemische Behandlung aus Eiweiss immer Benzoësäure und Derivate derselben gewonnen werden können, so hat für uns die Synthese der aromatische Kohlenwasserstoffe noch besonderes Interesse.

"Bei der zerstörenden Wirkung der Hitze aus einer grossen Zahl selbst der allereinfachsten Körper der kohlenstoffärmeren Classe von Verbindungen, entstehen Substanzen, welche der durch höheren Kohlenstoffgehalt ausgezeichneten Körperklasse angehören. Kekulé sagt, dass die Hitze den Kohlenstoff zu solch dichterer Aneinanderlagerung geneigt macht.

"Bei den hohen Hitzegraden entstehen die einfachsten aromatischen Verbindungen, wie Benzol, Homologe etc., und bei noch höheren Hitzegraden das an Kohlenstoff noch reichere Naphthalin. Diese Thatsachen sind besonders durch Berthelot begründet.

"Man sieht, wie ganz ausserordentlich und merkwürdig uns alle Thatsachen der Chemie auf das Feuer hinweisen, als die Kraft, welche die Constituenten des Eiweisses durch Synthese erzeugt hat. Das Leben entstammt also dem Feuer und ist in seinen Grundbedingungen angelegt zu einer Zeit, wo die Erde noch ein glühender Feuerball war.

"Erwägt man nun die unermesslich langen Zeiträume, in denen sich die Abkühlung der Erdoberfläche unendlich langsam vollzog, so hatten das Cyan und die Verbindungen, die Cyan- und Kohlenwasserstoffe enthielten, alle Zeit und Gelegenheit, ihrer grossen Neigung zur Umsetzung und Bildung von Polymerieen in ausgedehntester und verschiedenster Weise zu folgen und unter Mitwirkung des Sauerstoffs und später des Wassers und der Salze in jenes selbstzersetzliche Eiweiss übergehen, das lebendige Materie ist.

"Ich glaube also, dass von der leblosen zur lebendigen Natur ein Zwischenstadium führt."

In Pflüger's theory the protein which results of chemical evolution is called "lebendiges Eiweiss", by which he does not mean that it is alive, but that it is able to release "*vis viva*" (our kinetic energy, called "living energy" by Clausius) owing to the presence, in its large molecules, of what we call today "high energy bonds", characterized by high "Spannkraft" (our potential energy).

According to Pflüger, as he states it, the intramolecular heat of the cell is life, and the essential feature of life is the presence, in the living protoplasm, of "lebendiges Eiweiss", in which, according to his theory, nitrogen is largely in the form of cyanogen, the CN radical being known as highly reactive.

Pflüger's theory of the origin of life did not survive the demise of his cyan-protein theory.

The first theory of the origin of life which gained wide acceptance was

References p. 259

Plate 1. Alexander Ivanovich Oparin.

formulated by Oparin in 1922. Until the end of the 19th century, photo-synthesis was believed to be the source of all organic matter. In line with this view the first organisms were thought to have been autotrophs. But once the extreme metabolic complexity of these organisms was under-stood and the extreme complexity of their morphological organization revealed, it became difficult to regard them as primitive. When the chemoautotrophs were discovered, they in turn were thought to have been the first organisms to inhabit our planet; this view continued to be held for some time by a number of authors [19–21]. Yet the chemoautotrophs are of great metabolic complexity, and in 1922 Oparin[22] suggested that the first living organisms to develop on earth were able to nourish themselves heterotrophically on organic substances, the latter presumably having arisen abiogenetically long before the appearance of life on earth. In his 1922 note Oparin formulated the important theory concerning the self-assembly of abiogenetically formed organic molecules, producing the first precursors of cells. This view, which corresponds to the most extreme version of heterotrophy, stems from Mendeleev's theory concerning the inorganic origin of petroleum and from knowledge concerning the existence of organic substances in meteorites*.

The concept of organochemical evolution radically altered contemporary thinking on the subject of life's origins. It was the subject of Oparin's first pamphlet (1924) on the origin of life [23, 24]. A second and more elaborate book was published by Oparin[25] in 1936 and translated into English in 1938[26]. Oparin's new theory gained wide acceptance and entirely changed scientific thinking about the beginnings of chemical evolution, prior to the steps leading to living cells.

Oparin's theory concerns the genesis of carbonaceous molecules through the interaction of water and metallic carbides formed in the iron core of the earth. At the time, the prevailing view was that the starting temperature of the earth was very high; in Oparin's view hydrocarbons were the first to be formed. This is based on experimental evidence and he formulates his theory as follows:

"if we treat carbides of metals with superheated steam we obtain what are known as hydrocarbons, that is to say compounds consisting of carbon and hydrogen. These com-pounds must also have arisen when the carbides and steam met on the surface of the earth. Of course some of these must immediately have been burnt, being oxidized by the oxygen of the air. However, under the conditions then prevailing, this combustion must

---

* Academician Oparin's personal communication to the author.

have been far from complete. Only a certain part (and a comparatively small one) of the hydrocarbons were fully oxidized, being converted to carbonic acid and water. A further part, owing to incomplete oxidation, gave rise to carbon monoxide and oxygen derivatives of hydrocarbons, while finally, a certain proportion of the hydrocarbons, completely escaped oxidation and was given off into the upper, cooler layers of the atmosphere without any alteration. The more the earth cooled, the lower became the temperature at which the interaction between the carbides and the water vapour took place, and less carbonic acid and more unoxidized hydrocarbons were formed." (from the translation by A. Synge[24])

The notion that the hydrocarbons were the first "organic" compounds on earth is substantiated, Oparin believes, by astrophysical data. According to the then current theory concerning the origin of the earth, Oparin considers that it has passed through the stages of yellow star and of red star. Spectroscopic studies on red stars showed that their atmospheres contain hydrocarbons. Furthermore, nitrogen carbide (now formulated C_2N_2) was found in the tails of comets, as well as CO.

"Thus", Oparin writes, "we can demonstrate beyond doubt the presence of hydrocarbons on a number of heavenly bodies. This fact gives full support to the conclusion we have already drawn. There came a time in the life of the earth at which the carbon which had been set free from its combination with metal and had combined with hydrogen formed a number of hydrocarbons. These were the first "organic" compounds on the earth.

"Although only two elements, carbon and hydrogen, enter into the composition of these compounds, these elements can join together in the most varied combinations and give rise to the most varied hydrocarbons. Organic chemists can now list a very large number of such compounds." (translated by A. Synge[24])

Oparin concludes from the conditions of formation of these hydrocarbons that they were unsaturated molecules and that, if they could avoid oxidation at the time of their formation, they must, in the wet and hot atmosphere of the earth,

"have combined with oxygen and given rise to the most varied substances composed of carbon, hydrogen and oxygen in various proportions (alcohols, aldehydes, ketones and organic acids)." (translation by A. Synge[24])

Thus he concludes that at least a major portion of the carbon first appeared on the earth not in the form of CO_2 but in that of unstable organic compounds. Oparin traces the origin of nitrogen compounds to cyanogen formed by the reaction of the metallic nitrites with carbon compounds (hydrocarbons). When, as the theory asserts, the earth became a dark planet, its temperature fell to 100°C, enabling water to exist in the form of liquid drops. Rain fell on the earth and the first organic substances formed in the atmosphere fell into the primordial boiling ocean.

While still in the atmosphere, these substances had begun to react with one another. They combined with oxygen and ammonia, yielding compounds of hydrocarbons with oxygen or nitrogen. They continued to react in the boiling ocean and larger particles were formed. This hypothesis is based on experimental evidence and on the fact that if hydrocarbons are submitted to the conditions described, they are oxidized at the expense of oxygen and yield a variety of compounds (alcohols, aldehydes, acids, etc.), especially at high temperature and in the presence of iron or other metals. Oxidized hydrocarbons in reacting with one another produce more complicated compounds and they can also combine with ammonia, thus producing various nitrogen derivatives.

The great impact of Oparin's pamphlet was owing to its convincing formulation of the concept of *prebiological organo-chemical evolution*. Of course the details of his theory were modified with time, in particular by himself in subsequent publications[25, 26], as knowledge progressed. But there is no doubt that Oparin's name must be attached to the acceptance of the theory that the beginnings of life on earth were preceded by abiogenic molecular (chemical) evolution. First formulated in 1922 and developed in 1924 in a monograph of broad conceptual systematization, the theory became the general working hypothesis in the field.

In 1929, Haldane[27] in a general article in *"The Rationalist Annual"*, wrote the following sentences:

"Now, when ultra-violet light acts on a mixture of water, carbon dioxide and ammonia, a vast variety of organic substances are made, including sugars and apparently some of the materials from which proteins are built up. This fact has been demonstrated in the laboratory by Baly of Liverpool and his colleagues. In this present world, such substances, if left about, decay — that is to say, they are destroyed by microorganisms. But before the origin of life they must have accumulated till the primitive oceans reached the consistency of hot dilute soup."

This phrase expresses the concept of chemical evolution. It also contains the idea of an atmosphere originally deprived of oxygen, from which he draws the conclusion that at the time ultraviolet light could, due to the lack of the ozone sheet, reach the earth from the sun. The aphoristic statement is as remarkable as it is consequential. It also contains the less felicitous notion of the "hot dilute soup", which became very popular.

It may be noted that the interpretation of the experiments done by Baly, on which Haldane based his argument, are not accepted today (see Stiles[27a]). Likewise, the concept of a reducing primitive atmosphere has been abandoned. Despite its great popularity, the "hot dilute soup" teeming with organic molecules is a misrepresentation which should be abandoned (see Florkin[28]). The "hot soup" is a closed system, and it must be supposed that the collection of open systems of the biochemical continuum of today derived from an open prebiological continuum.

3. The primitive atmosphere

It is now generally believed that during the formation of the earth through the accretion of materials dispersed in a cold cloud of dust or gas, the gaseous phase of the forming earth was lost and replaced by an atmosphere coming from the interior of the earth (primitive secondary atmosphere) and containing only trace amounts of molecular oxygen. On this point the geological and thermodynamic evidence was thought to confirm the view formulated by Haldane concerning a primitive reducing atmosphere. From a number of different types of evidence Urey[29] concluded that carbon and nitrogen were present in the primitive atmosphere, largely in the form of methane and ammonia. Another generally accepted kind of evidence comes from the knowledge that the *Jovian* planets (Jupiter, Saturn, Uranus, Neptune) have been shown by spectroscopic analysis to be surrounded by planetary atmospheres whose main components are methane and ammonia. In 1959, Urey and his collaborator Miller[30] expressed the view that the primitive atmosphere of the earth as well as of the other terrestrial planets (Mars, Venus, Mercury) contained CH_4, NH_3 and H_2 and were highly reduced.

Miller's experiment was one of the first to be carried out with a view to analyse the possible realisation of one particular stage of the hypothetical chemical evolution. Miller[31], in 1953, simulated the primitive secondary atmosphere (methane, ammonia, hydrogen, water vapour at a pressure of 1 atm.) in a closed-glass apparatus, subjected this simulated atmosphere to electric discharge for a week and found in the resulting reaction mixture, by means of paper chromatography, several α-amino acids characteristic of protein. This experiment, though not now considered geologically adequate, was of considerable influence and opened up a new field of investigation.

4. Synthesis of monomers and of polymers

Experimental data have been collected in great numbers concerning the possible modes of synthesizing the various classes of components under

primitive earth conditions. In experiments of this kind, the probable constituents of the primitive atmosphere, which were at the time thought
predominantly to consist of a source of carbon, CH_4, a source of nitrogen
NH_3, and a source of oxygen, H_2O, are exposed, in appropriate equipment,
to sources of free energy such as ultraviolet rays, electric discharges, heat
or ionizing radiations. Amino acids have been synthesized through the
application of these free-energy sources under simulated primitive earth conditions. In some cases, the initial material was composed of reactive intermediates obtained in another type of condition. Ponnamperuma and
Gabel[32] as well as Keosian[33] and Kenyon and Steinman[34] have recorded
the history of this work and the reader is referred to their writings.

But the mechanism of these syntheses of amino acids remains unknown in
many cases. Miller[31] suggested two possible mechanisms for his experiment.
First, aldehydes and hydrogen cyanide are synthesized through the spark
(electrons and heat) in the gas phase. In the aqueous phase, they react and
produce amino and hydroxynitriles, which in turn are hydrolyzed into
amino acids and hydroxyacids. The second mechanism he suggested is the
formation, through the spark from the ions and radicals produced, of
amino and hydroxy acids.

Oro[35] has shown that adenine could be synthesized from a concentrated
solution of ammonium cyanide, and this was confirmed by Lowe et al.[36].
Oro[37] has made a detailed study of the course of the reaction and shown
that the probable intermediates were 4-aminoimidazole 5-carboximide and
formamidine. Ponnamperuma et al.[38] also synthesized adenine by the electron irradiation of a mixture of methane, ammonium and water, the
yield being about 0.01 per cent of the methane at the start. In the synthesis
by electron irradiation, aminomalononitrile is an intermediate[39]. The action
of electric discharges on a mixture of methane and hydrogen produces as the
major nitrogen-containing product cyanoacetylene[40] which can act as a
starting point for a number of pyrimidine syntheses. According to Oro's
theoretical views[41], the synthesis of pyrimidine could start from C-3 molecular species found in comets, one of these compounds being malon-
amidesemialdimine or its isomer β-aminoacrylimide.

One of the most striking experiments in the field was conducted by
Fox and Harada[42]. While the heating of an amino acid results in its degradation, in the presence of aspartic acid or glutamic acid such heating leads
to the formation of copolymeric peptides.

These are a few examples of the numerous experiments carried out with the components believed to have been present in the primitive atmosphere and in the conditions prevailing there (literature in Ponnamperuma and Gabel[32] and in Kenyon and Steinman[34]).

The generally accepted scheme of chemical evolution begins with the application of energy (electric discharges, UV rays, heat, ionizing radiation) to compounds believed to have been available in the primitive atmosphere, such as CH_4, NH_3, CO, H_2O, H_2, etc. Under the action of these free-energy sources, these substances were transformed into reactive compounds such as nitriles (R–CN) and aldehydes (R–CHO) which were brought by the rain into the primitive ocean; there, they were subsequently transformed into such biomonomers as amino acids, heterocyclic bases, carbohydrates, fatty acids, porphins, etc. and polymers such as polypeptides, polynucleotides, etc., from which the first primitive cells emerged. These views revive in a striking manner, and in a new context, the theory which held methane, ammonia, water and hydrogen to be the "types" from which all organic compounds could be derived (see chapter 12 of Vol. 30). According to the "theory of types", all organic compounds can be derived from inorganic "types". As stated in Chapter 12 of Vol. 30, the first "type" proposed by Hofmann in 1849 was ammonia. The second, water, was proposed by Williamson in 1850. Gerhardt, in 1853, held that the organic compounds could be derived from four types: H_2O, NH_3, H_2 and HCl, to which Kekulé added the "type" CH_4 in 1856 (literature in Partington[43]). The attention given to the "types" was questioned by Kekulé himself in 1858; he then attached greater importance to a study of the relation between the properties of the elements and those of their compounds. As a consequence the theory of "types" was discarded.

A new viewpoint (preenzymatic metabolism) on chemical evolution was introduced by Buvet[44] who individualizes in metabolism (as we see it presently) the implication of elementary types of reactions which he considers as having existed, in aqueous media, before enzymes.

5. The proteinoid theory of the origin of life

In the classical analogy between the organism and the machine, it is commonly assumed that neither could spontaneously arise from a mixture of their component molecules. This was applied to the different levels of structure, and when the orderliness of protein macromolecules was revealed, the same concept was applied to them, as the progress of molecular biology had revealed the cellular mechanism of their replication. Nevertheless, in contradistinction to the current dogma of biochemical genetics, a first type of self-organization was revealed when Schmitt[45] showed, as stated in Chapter I (Vol. 29A), that if the collagen of a fish swimbladder is dissolved in aqueous acetic acid solution and precipitated by dialysis, the protein molecules assemble in regular fibrils with periodic cross striation. Other examples of self-association have been discovered since (see Chapter I, Vol. 29A).

In the field of chemical evolution the concept of self-organization of matter into cells had been raised by Oparin in his 1924 pamphlet and in his subsequent books[23, 25]; this led him to experiment with coacervate droplets. As Fox has emphasized, the coacervate droplet "fails crucially to explain the origin of the first cell". At the time of Oparin's publication, the charge of information carried by such biomolecules as gelatin and other biopolymers used to prepare the coacervates was unknown and Fox rightly states "the fundamental question is one of how cells would emerge from polymers which were *not* the product of cellular synthesis". This question was resolved in the work Fox and his collaborators did on the abiotic polymerization and self-organization of amino acids in the form of "proteinoids". This line of research stemmed from an interest in the nature of primoridial protein on earth. Like Blum[46, 47], Fox realized that an explanation should be given for the formation of peptide bonds:

"When we first began to wonder about the terrestrial origins of proteins, two considerations inhibited investigative action. Although it seemed likely that peptide bonds must have formed originally at elevated temperatures in order to split out water molecules, we supposed that such temperatures would cause amino acids to "go up in smoke". The second consideration was also a prejudgment — the difficulty of visualizing how thermal conditions might provide protein products with what we take to be a necessary specific arrangement of amino acid residues." (ref. 48)

Other experiments suggested that

"in a preenzymic world, amino acids reacting at elevated temperatures might determine their own order in a primordial peptide chain." (ref. 48)

As a result, the heating procedure was tried i an initially dried state so as to overcome the energetic barrier which exists in aqueous solution, the temperature being high enough to distil water and to polymerize the dried residue. The view that amino acids determine their own order in a chain was derived from studies of papain-catalyzed syntheses of acetylpeptide-

Plate 2. Sidney W. Fox.

anilines[49]. In these studies, glycinanilide was tested with each of thirteen benzoylamino acids only one of which participated: benzoylglycine (hippuric acid). Fox and his colleagues have shown that it is possible to obtain polymers of amino acids simply by heating initially dry mixtures of amino acids at temperatures higher than 100 °C (literature in Fox and Dose[50]). The term *proteinoid* coined by Fox and Harada[42] has come to mean

"macromolecular preparations of mean molecular weights in the thousands and, qualitatively, contents of most of the eighteen amino acids found in proteins. While these polymers have other properties of contemporary proteins as well, identity with the latter is not a necessary inference." (ref. 42)

The preparation of proteinoids may appear to refute the proposition by Carothers[52], the inventor of nylon, that α-amino acids could not be polymerized thermally. This is true of most amino acids (the "neutral" ones) but the evidence obtained from the experiments conducted by Fox and his collaborators goes beyond Carother's inference in showing that amino acid mixtures containing basic amino acids or acidic amino acids can be copolymerized by heat, to include all of the proteinaceous amino acids.

One of the most interesting features of proteinoids is that they are bioseme carriers[53], as revealed by selective interactions with enzyme substrates[54] and selective interactions with polynucleotides[55]. A large amount of work on proteinoids has been performed in a number of laboratories (literature in Fox and Dose[50]). As each kind of proteinoid, due to the ordering of amino acids in their molecules, has its array of weak enzyme-like activities, though feebly and with little specificity, the thermal proteinoids afford an answer to the query of how enzymes came into existence when there were no enzymes to make them. The enzymatic activities of proteinoids has been the subject of study in different laboratories. These activities include pH–activity curves of the usual type, Michaelis–Menten kinetics, heat inactivation in aqueous solutions, specificity of interaction between proteinoid and substrate, etc. (literature in ref. 50). The striking properties of proteinoids have made the thermal theory of the origin of proteins very attractive; as a result, the older experiments performed with electric discharge (the spark consisting of electrons and heat) are no longer the centre of attention. We showed above how Carothers' proposition came to be discarded. The experiments conducted by Fox and his collaborators are also geologically relevant unlike others such as that by Miller, carried out with a mixture of gases containing 75 per cent H_2, and many other experiments conducted in closed systems.

Other theories concerning the abiotic formation of protein-like polymers have been formulated but none of them has led to the preparation of such interesting copolymers as the thermal proteinoids.

Given our contemporary familiarity with phenomena of self-assembly which have been recognized at the level of cell organelles, etc. (see Chapter I, Vol. 29A), we have no difficulty in imagining how polyaminoacids could assemble to form protocells. One obstacle that remained was the question of the source of information. The presence of information in the heat proteinoids indicates the possibility of self-assembly of amino acids, with information coming from the diverse reactant amino acids. This is geologically relevant as amino acids may have been formed from CH_4, NH_3 and water vapour present in hot gas from volcanoes; they may have undergone heat polymerization into proteinoids in the lava or in volcanic ash.

When the proteinoids enter into contact with water as may have happened when they were extracted by rainfall from the ash and brought to the surface waters of the earth, they produce vast numbers of microsystems. When hot solutions of thermal proteinoids are allowed to cool slowly (for one or two weeks) in fixed conditions of pH and salt concentration, spherical droplets about 2 μ in diameter are formed (microspheres). Microspheres cannot be compared with the many simulations of organic forms which are found in the old literature. They are made of proteinoids and thus of great interest because of their ordering and catalytic properties (for the literature on microspheres, see Fox and Dose[50]). In addition, they shrink in response to hypertonic solution and swell in hypotonic solution; they can be made to stain Gram-positive or Gram-negative; they manifest a degree of stability greater than that of cells and they show a kind of replication[53]. Such microsystems also possess an inherent tendency to communicate[56] and when they join, smaller particles composed of proteinoids are transferred from one microsphere to another.

Various aspects of the theory of proteinoids are discussed by Fox and Dose[50].

Fox considers the proteinoid microsphere a model for "the primitive cell possessing a number of salient properties of contemporary cells". Fig. 1 represents the flow chart of principal stages of molecular evolution and the origin of life.

The flow chart of Fig. 1 starts with amino acids in the environment. Their presence can be explained by the reactions involved in the many experiments conducted since Miller's first publication on the subject. However, a new trend of thought was introduced in 1968 and 1969 by the discovery (made possible by microwave spectroscopy) of interstellar organic matter (Cheung et al.[57], Snyder et al.[57a]). This discovery revealed the presence of formaldehyde, ammonia, hydrocyanic acid and water at different points in

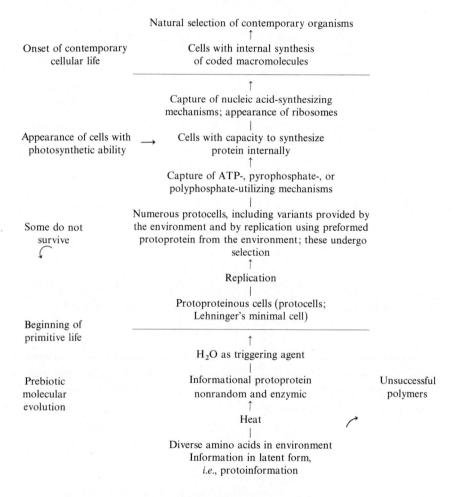

Fig. 1. Flow chart of principal stages of molecular evolution and the origin of life. (Fox and Dose[50])

References p. 259

our galaxy. These compounds have low densities and low kinetic temperatures but they may have been present in clouds where physical condensation could have taken place, though it is difficult to assess the significance of this theory of the origin of amino acids on earth.

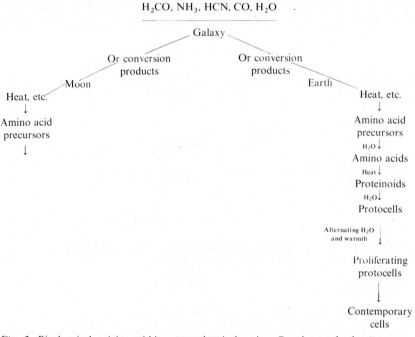

Fig. 2. Biochemical origins within cosmochemical unity. On the earth the flow may have proceeded as indicated on the right; on the moon it may have terminated as indicated on the left. (Fox and Dose[50])

Taking into consideration the new data on interstellar organic matter, Fox[58, 50] draws a chart of biochemical origins within cosmochemical unity (Fig. 2). The proteinoid theory of the origin of life implies a succession of steps involving heat and water.

$$
\begin{array}{c}
\text{amino acids} \\
\downarrow \text{HEAT} \\
\text{protein(oid)s} \\
\downarrow \text{WATER} \\
\text{microspheres}
\end{array}
$$

These two steps may have taken place in different regions of the earth as stated above.

The attraction of the proteinoid theory is that in its constructive aspect it reveals unexpected properties and reactions.

Fox and Dose[50] list six major questions to which they consider the theory gives an answer:

"*The origin of order in proteins when no macromolecules and no contemporary code existed.*" The experiments underlying the proteinoid theory show that the information was provided by the monomers for the formation of the first protein (on amino acid side-chains as structural biosemes, see Chapter I, Vol. 29 A).

"*The origin of enzymes when no enzymes to make them existed.*" Proteinoids have the characteristics of some enzymes, depending on the mixture of amino acids used and on geophysical conditions.

"*The origin of metabolism in the absence of metabolizing cells.*" The origin of metabolism is explained by the association of individual reactions catalyzed by proteinoids.

"*The origin of cells when no cells existed to produce them.*" The polymeric combinations of amino acids, including those produced by heat, assemble themselves into microsystems.

"*The origin of membranes when no microsystems containing membranes existed.*" Proteinoid microspheres display double layers permitting the retention of large molecules while allowing smaller molecules to diffuse out.

"*The origin of reproduction.*" Proteinoid microspheres participate in the reproduction of their own likeness.

A number of scientists are reluctant to accept the proteinoid theory. In his recent book *L'Origine des Êtres Vivants et des Processus Biologiques*, Buvet[44] expresses the view that the proteinoid microspheres obtained their constituents from an altogether too unlikely origin. The data obtained by Oparin's school are also subject to his criticism (as they were to that of Fox) that the compounds utilized have not been produced in abiogenetic experiments.

'The proteinoid–microsphere theory seems too simplistic to many scientists. Logically it is not impossible for the heating phase to have occurred on the slope of a volcano and for rain to have washed out its result in the form of microspheres. But you will never make the "hot dilute soup" enthusiasts swallow this concept. The present writer believes that the real virtue of the proteinoid–microsphere theory resides in the possibility of obtaining microspheres from mixtures of amino acids and in the unexpected properties such microspheres were proved to possess. To believe that things really happened as the theory says they did would certainly be simplistic and naive, but that does not invalidate the theory as a constructive line of thought nor does it render worthless experiments that have produced unexpected results. One of these is the production of informed proteins in the absence of the sequencing information provided in contemporary life by nucleic acids. Another way of getting polypeptides without nucleic acids has been described by Lipmann[59]. Another epistemological obstacle placed in the way of the proteinoid theory is the notion that the heating of amino acids yields disordered polymers. There is now extensive evidence, however, that thermal condensation of α-amino acids produces ordered polymers (see Dose and Rauchfuss[60]).

In the proteinoid theory, the nucleic acid mechanism appears as having arisen as an evolutionary refinement. One may suppose with Jukes[61] that preenzymatic proteins of a proteinoid nature have been incorporated into a spherical bileaflet boundary structure and that they have translated their amino acid sequence into ribonucleic acid through the intervention of RNA-polymerase protein. Such an RNA molecule could then be copied in a DNA structure following a method analogous to that used in hybridization experiments.

6. The gene hypothesis of the origin of life. Nucleic acids before proteins

In the proteinoid–microsphere theory outlined above, the nucleic acid mechanism appears as having arisen as an evolutionary refinement. However, several authors in complete opposition to the views derived from Oparin's 1922 theory, maintained that nucleic acids must have preceded proteins in the history of life. According to Miller and Horowitz[62], for instance, the first living organisms were composed of a polynucleotide associated with a polymerase.

The most modern version of this line of thought (Crick[63], Orgel[64]), opposing the question of life without proteins to that of life without nucleic acids, veers towards explaining the formation of the first proteins in the absence of enzymes by the self-assembly structure of a primitive ribosome entirely composed of RNA. The theory is derived from modern biochemical genetics and has no experimental basis. We therefore refer the reader to the relevant literature[63, 64].

The accumulation of experimental data related to chemical evolution shows that the nucleotide bases, the amino acids, the carbohydrates as well as the polypeptides and polynucleotides may in prebiotic conditions have been synthesized, albeit inefficiently, in the absence of biocatalysts.

Orgel[64] in a thoughtful essay suggests

"a crude correlation was established between certain polynucleotide sequences (codons) and certain available amino acids (or groups of amino acids) and that the whole system subsequently perfected itself by the bootstrap principle..."

While he accepts the theory concerning the development of life in the absence of nucleic acids so far as the formation of polymers of amino acids is concerned, he does not conceive that a system of proteins without nucleic acids could have resulted in life as we know it.

"No great progress" — he writes — "would have been possible in the absence of a fairly accurate form of residue-by-residue replication".

It may be noted here that such a statement does not contradict the theories of protein formation before the advent of a spontaneously developed unit-by-unit replication of polypeptides.

To support the theory that life evolved out of nucleic acids in the absence of proteins, Orgel cites the arguments in favour of the complementary replication, in the absence of enzymes and produced by certain structural characteristics inherent in the bases themselves, of nucleic acids (for literature, see Orgel[64]). But as we have seen and as Fox has shown, such a character is also observed where the self-association of amino acids results.

Orgel entitles his book on the subject: *The Origins of Life — Molecules and Natural Selection*[65]. He distinguishes between informed polypeptides (determined by a preformed polynucleotide) and non-informed polypeptides which could have accumulated before the first nucleic acids. Natural selection is conceived as operating on the informed polypeptides only.

"A system containing nucleic acids and informed polypeptides is subject to natural selection; those nucleic acids that direct the formation of useful informed polypeptides are successful in eliminating their less "talented" competitors."

In Orgel's conception, natural selection (conceived by him as active on the macromolecular level) could not act on protein sequences until nucleic acid replication was underway.

He recognizes the possibility of a catalysis of the formation and replication of the first nucleic acids by uninformed polypeptides that could have existed since the beginning.

References p. 259

7. The thermodynamics of the irreversible processes and the origin of life

The energetics of evolving systems have lately been developing in new directions.

The energy received by the surface of the earth in the form of solar radiation is dissipated as heat in outer space. In the case of living bodies, this process takes a special form. Part of the energy accumulated by photosynthesis in the form of organic compounds is metabolized through chains involving electron transfers and accumulated in the form of the energy-rich bonds of ATP, from which it is used for the performance of work and finally dissipated in the form of heat in outer space. The biosphere composed of the mass of organisms and their media is in a steady state removed from the equilibrium state and in its whole mass the entropy increases with time. Schrödinger[66] in an influential book published in 1948 noted that when one considers the organisms themselves and their mechanisms of hereditary transfer one must conclude that in the succession of organisms a particular order is maintained. In the whole of the biosphere, entropy increases with time but it is kept at lower levels at certain points of the system by the change-over from randomness to order. Thus Schrödinger spoke of the organism as "sucking orderliness from the environment", of feeding on "negative entropy" (abbreviated to negentropy by Brillouin). The relation between the theory of information and the concept of entropy was worked out by Shannon and Weaver[67] and by Brillouin[68], after Szilard[69] had recognized the formal similarity between the mathematical expressions of entropy and of information which is expressed as the negative values of entropy. Fig. 21 (Vol. 29A, Chapter I, p. 69) illustrates the increase of information which took place in the evolution of organisms.

Until around 1930, the second principle continued to underlie the theories of physical and biological evolution. When the present author published[70] in 1943, a handbook of biochemistry reviewing our basic views on metabolism, Th. de Donder, Professor at Brussels University and a leading scholar in the field of thermodynamics expressed the opinion that complicated interpretations of metabolic pathways told us nothing we did not already know from the second law of thermodynamics. It is nevertheless the school inspired by this great scientist which changed our views in that field by the development of the thermodynamics of irreversible processes. This type of theory gained wide acceptance particularly after the publication, by two disciples of De Donder, Glansdorff and Prigogine[71], of a macroscopical criterion for the evolution of open systems, in a state remote from equilibrium.

As stated by Buvet[44]:

"This criterion constitutes in fact a generalization of the theorem of minimal entropy production. It expresses that in the course of the evolution of a system maintained permanently in exchange regime, part of the entropy production bound to the variations of thermodynamical forces: temperature, pressure, concentration gradients, reaction affinities, etc. can only decrease. Near equilibrium, this criterion is reduced to the theorem of minimal entropy production. Far from equilibrium, when the relations between flux and forces are no longer approximately linear, the criterion introduces new transition properties." (translation by author)

Since 1965, examples of structure production by the flux of energy through a system have been described. Examples may be found in the experiments on Knudsen's membranes, Bénard's rings, etc. (see Glansdorff and Prigogine[72]).

To quote from Vol. 29 A, Chapter I (by the present author) of this Treatise:

"Short time oscillations have been detected in a number of biochemical reactions and arguments have been formulated in favour of instabilities breaking dissipation symmetry and leading to spatial organization."

Eigen[73] has formulated a theory of molecular selection based on the combined utilization of nucleic acids and of proteins in the selection leading to a reproductive process. Nucleic acids provide complementary instruction without significantly contributing to catalytic activity. It is the reverse for proteins presenting an enormous functional and recognitive diversity and specificity.

"Via catalytic couplings they may link together many information carriers and thus build up a very large information capacity."

The origin of life is seen by Eigen as the nucleation of the functional correlation by which "information" acquires its meaning.

"Any fluctuation in the presence of potential coupling factors leading to a unique translation, and its reinforcement *via* the formation of a catalytic hypercycle, offers an enormous selective advantage and causes the breakdown of the former steady state of uncorrelated self-reproduction."

References p. 259

Eigen illustrates the concept of "catalytic hypercycle" by a model made of an *ensemble* of nucleic acids and proteins organizing itself into a stable, self-reproducing and evolving unit. The model consists of short nucleotide sequences (I_i) providing information for one or two catalytically active polypeptide chains (E_i) (see Eigen[73]). This model belongs to the type of theory that starts from the present biological situation. Fox[74] has noted that the constructionist (going from simple to complex) theory of proteinoid microspheres also meets the requirements of Eigen's catalytic function in combination with a feedback mechanism ("with observation of the hetero-trophic replicability at the system level and the concept of continuously generated enzyme-like activities at the macromolecular level...").

It appears therefore that the extensive theories on the thermodynamics of open systems maintained at a state remote from equilibrium give us an interpretation of the quantitative characters of chemical prebiological evolution and of biochemical evolution, which appears as inevitable and sub-mitted to the statistical causality of such systems (see Schoffeniels[75]).

Concerning the process of chemical prebiological evolution as well as of biochemical evolution, the interpretations cannot, as Eigen emphasizes, derive from the extensive thermodynamical aspect but must derive rather from nucleated intensive aspects belonging to the domain of molecular biosemiotics, as described in Chapter I (Vol. 29 A).

8. Epistemological aspects

If it is true that, generally speaking, science is the science of an object which is not history, yet in the particular case we are now dealing with the object *is* history, *i.e.*, the historical steps which preceded the appearance of the first living cells. We are writing a history within history, that is, the history *in human thought* of the history of living systems.

Consequently, the data assembled under the heading "chemical evolution" deserve this predicate only if they lead to proliferating systems.

On the other hand, experimentation in the field is governed only by mental attitudes and by long-range prevision as to the approach to the historical steps which have led to protocells.

As stated above, the aspects considered in this chapter do not directly concern the study of the origin of life but the gathering of a series of ex-perimental data which may have bearing on the theory concerning the origin of life (experimental prebiology). These data are collected in the course of application of high-energy sources to simple reactants such as N_2, NH_3, CO. CH_4, H_2O and H_2; underlying such application is the theory of types, current in the mid-19th century, according to which the many forms of organic compounds could be derived from such elementary types. The overall conclusions drawn from these experiments tend to suggest the formation of nitriles and aldehydes as reactive compounds from which such monomers as amino acids, heterocyclic bases, carbohydrates, fatty acids, porphins, etc. may be obtained. From these monomers such polymers as polypeptides, polynucleotides, etc. may in turn be derived. Of course such experiments can only be of interest if they are geologically relevant. First the primitive atmosphere was considered either highly reduced or predominantly composed of methane, ammonia, water and hydrogen. Nowadays, it is believed that the major portion consisting of hydrogen, methane and ammonia was lost and that the — less reducing — atmosphere as it existed 3.5 billions years ago was composed mainly of nitrogen, hydrogen, carbon monoxide and carbon dioxide with small amounts of methane and ammonia. However, there are many experiments showing that amino acids and a number of organic molecules may be derived, through HCN, in the irradiation of mixtures of CO_2, CO, H_2 and N_2.

It has also been claimed that the mean temperature was not above 100 °C but what counts is the temperature at a definite spot, which may be much higher.

It is no surprise, in an endeavour which is imaginative in character and tries to reconstitute not actual events but tries, by a recourse to simulation by experiment, to find the presence of a number of themes conceived by the spontaneous exercise of the structure of the human mind as well as concepts which have resulted from human scientific enquiry on life as we know it today. To the latter belongs the reference to the data of biochemical genetics.

Recent developments in the theory of the thermodynamics of irreversible processes have opened up new vistas in the field of extensive generalized concepts relating to the origin of life and brought these concepts back to the statistical causality of universal natural laws. This domain is in the field of application of the second law of thermodynamics from which the evolution criterion established by Glansdorff and Prigogine is a consequence. As stated by Buvet[44]:

"Introducing the degradation of energy, this principle would carry in itself its own moderation, and the expression of this moderation would coincide with what we are used to calling
LIFE, lacking so far a better definition of it, as a consequence of the necessary complexity
of the systems endowed with this property."

Extensive thermodynamical concepts of the theory of open systems, crossed
by an energy flux and and maintained in a state remote from equilibrium
appear suitable for application to the system of life and seem to point to the
inevitable molecular character of evolution. As stressed by Eigen, the nature
of the process of those aspects of molecular evolution can only be grasped
at the level of the intensive properties which we have, in Chapter I
(Vol 29 A) brought together under the heading "*molecular biosemiotics*".

REFERENCES

1 J. B. van Helmont, *Ortus Medicinae*, Amsterdam, 1667.
2 F. Redi, *Experiments on the Generation of Insects (1688)*, transl. M. Bigelow, Chicago, 1909.
3 For literature, see W. Bulloch, *The History of Bacteriology*, London, 1936.
4 L. Joblot, *Observations d'Histoire Naturelle*, 2 Vols., Paris, 1754–1755.
5 J. T. Needham, *Phil. Trans. Roy. Soc. (London)*, No. 490 (1749) 615.
6 L. Oken, *Die Zeugung*, Bamberg, 1805.
7 N. Appert, *L'Art de Conserver, Pendant Plusieurs Années, Toutes les Substances Animals et Végétales*, Paris, 1810.
8 L. J. Gay-Lussac, *Ann. Chim. Phys.*, 76 (1810) 245.
9 C. F. Burdach, *Die Physiologie als Erfahrungswissenschaft*, 2nd edn., Leipzig, 1832.
10 H. M. Edwards, *Leçons sur la Physiologie*, 14 Vols., Paris, 1857–1881.
11 J. Théodoridès, *Stendhal du côté de la Science*, Aran (Switzerland), 1972.
12 L. Pasteur, *Ann. Phys. (Paris)*, 64 (1862) 184.
13 F. Pouchet, *Hétérogénie ou Traité de la Génération Spontanée basée sur de Nouvelles Expériences*, Paris, Baillière, 1859.
14 J. Tyndall, *Essays on the Floating Matter in the Air in Relation to Putrefaction and Infection*, London, 1881.
15 J. Farley, *J. Hist. Biol.*, 5 (1972) 85.
16 S. W. Fox, *Encycl. Polymer Sci. Technol.*, 9 (1968) 284.
17 *Nouveau Larousse Illustré*, Paris, 1896–1904.
18 H. Mondor, *Pasteur*, Paris.
18a W. Pflüger, *Arch. ges. Physiol.*, 10 (1875) 251.
19 C. H. Werkman and H. G. Wood, *Advan. Enzymol.*, 2 (1942) 135.
20 M. Stephenson, *Bacterial Metabolism*, 3rd edn., London, 1949.
21 W. O. Kermack and H. Lees, *Sci. Prog. (London)*, 40 (1954) 44.
22 A. I. Oparin, Communication to the Meeting of the Russian Botanical Society, Moscow, 1922.
23 A. I. Oparin, *Proiskhozdenie Zhizny*, Moscow, 1924.
24 Translation of ref. 23 by A. Synge, in: J. D. Bernal, *The Origin of Life*, London, 1967.
25 A. I. Oparin, *The Origin of Life*, 1st edn. Translation by S. Morgulis, New York, 1938; 2nd edn., 1941.
26 A. I. Oparin, *The Origin of Life on the Earth*, 3rd edn., revised and enlarged. Translation by Ann Synge, Edinburgh and London, 1957.
27 J. B. S. Haldane, *Rationalist Annual*, (1929) 148.
27a W. Stiles, *Photosynthesis*, London, 1925.
28 M. Florkin, *Bull. Cl. Sci., Acad. Roy. Belg.*, 55 (1969) 257.
29 H. Urey, *The Planets*, New Haven, Conn., 1952.
30 S. L. Miller and H. Urey, *Science*, 130 (1959) 245.
31 S. L. Miller, *Science*, 117 (1953) 528.
32 C. Ponnamperuma and N. W. Gabel, *Space Life Sciences*, 1 (1968) 64.
33 J. Keosian, *The Origin of Life*, 2nd edn., New York, 1968.
34 D. H. Kenyon and G. Steinman, *Biochemical Predestination*, New York, 1969.
35 J. Oro, *Biochem. Biophys. Res. Commun.*, 2 (1960) 407.
36 C. V. Lowe, M. W. Rees and R. Markham, *Nature*, 199 (1963) 219.
37 J. Oro, *Federation Proc.*, 20 (1961) 352.
38 C. Ponnamperuma, R. M. Lemmon, R. Mariner and M. Calvin, *Proc. Natl. Acad. Sci. (U.S.)*, 49 (1963) 737.

39   R. A. Sanchez, J. Ferris and J. E. Orgel, *Science*, 153 (1966) 72.
40   R. A. Sanchez, J. P. Ferris and J. E. Orgel, *Science*, 154 (1966) 784.
41   J. Oro, *Ann. N.Y. Acad. Sci.*, 108 (1963) 464.
42   S. W. Fox and K. Harada, *Science*, 128 (1958) 1214.
43   J. R. Partington, *A History of Chemistry*, Vol. 4, London, 1964.
44   R. Buvet, *L'Origine des Êtres Vivants et des Processus Biologiques*, Paris, 1973.
45   F. O. Schmitt, *Proc. Am. Phil. Soc.*, 100 (1956) 476.
46   H. F. Blum, *Am. Scientist*, 43 (1955) 595.
47   H. F. Blum, *Time's Arrow and Evolution*, Princeton, 1955.
48   S. W. Fox, *Am. Scientist*, 44 (1956) 347.
49   S. W. Fox, M. Winitz and C. W. Pettinga, *J. Am. Chem. Soc.*, 75 (1953) 5539.
50   S. W. Fox and K. Dose, *Molecular Evolution and the Origin of Life*, San Francisco, 1972.
51   T. Hayakawa, C. R. Windson and S. W. Fox, *Arch. Biochem. Biophys.*, 118 (1967) 265.
52   W. H. Carothers, *Trans. Faraday Soc.*, 32 (1936) 39.
53   S. W. Fox, *Naturwissenschaften*, 59 (1969) 1.
54   D. L. Rohlfing and S. W. Fox, *Advan. Catalysis*, 20 (1969) 373.
55   S. W. Fox, J. C. Lacey Jr. and T. Nakashima, in: D. W. Riblons and F. Woesmer (Eds.), *Nucleic Acid–Protein Interaction*, Amsterdam, 1971.
56   L. L. Hsu, S. Brooke and S. W. Fox, *Currents Mod. Biol.*, 4 (1971) 12.
57   A. C. Cheung, D. M. Rank, C. H. Townes, D. D. Thornton and W. J. Welch, *Phys. Rev. Letters*, 21 (1968) 1701.
57a L. E. Snyder, D. Buhl, B. Zucherman and P. Palmer, *Phys. Rev. Letters*, 22 (1969) 679.
58   S. W. Fox, *Ann. N.Y. Acad. Sci.*, 194 (1972) 71.
59   F. Lipmann, in: R. Buvet and C. Ponnamperuma (Eds.), *Chemical Evolution and the Origin of Life*, Amsterdam, 1970.
60   K. Dose and H. Rauchfuss, in: D. L. Rohlfing and A. I. Oparin (Eds.), *Molecular Evolution: Prebiological and Biological*, New York, 1972.
61   T. H. Jukes, *Molecules and Evolution*, New York, 1966.
62   S. L. Miller and N. H. Horowitz, in: G. S. Pittendrigh, W. Vischniac and J. P. T. Pearman (Eds.), *Biology and the Exploration of Mars*, Washington, 1966.
63   F. H. C. Crick, *J. Mol. Biol.*, 38 (1968) 367.
64   L. E. Orgel, *J. Mol. Biol.*, 38 (1968) 381.
65   L. E. Orgel, *The Origins of Life — Molecules and Natural Selection*, London, 1973.
66   E. Schrödinger, *What is Life?*, London, 1948.
67   C. Shannon and W. Weaver, *The Mathematical Theory of Communication*, Urbana, Ill., 1949.
68   L. Brillouin, *Science and Information Theory*, 2nd edn., New York, 1962.
69   L. Szilard, *Z. Physik*, 53 (1929) 840.
70   M. Florkin, *Introduction à la Biochimie Générale*, Paris, 1943.
71   P. Glansdorff and I. Prigogine, *Physica*, 30 (1964) 351.
72   P. Glansdorff and I. Prigogine, *Structure, Stabilité et Fluctuations*, Paris, 1971.
73   M. Eigen, *Naturwissenschaften*, 58 (1971) 465.
74   S. W. Fox, *Naturwissenschaften*, 60 (1973) 359.
75   E. Schoffeniels, *L'Anti-Hasard*, 2nd edn., Paris, 1974.

Subject Index